DRIVING HORSES

How to Harness, Align, and Hitch Your Horse for Work or Play

STEVE BOWERS AND
MARLEN STEWARD

Voyageur Press

First published as *Farming with Horses* in 2006 by Voyageur Press, a member of Quayside Publishing Group, 400 First Avenue North, Suite 400, Minneapolis, MN 55401 USA. This edition published 2014.

The information in this book is true and complete to the best of our knowledge. All recommendations are made without any guarantee on the part of the author or Publisher, who also disclaims any liability incurred in connection with the use of this data or specific details.

We recognize, further, that some words, model names, and designations mentioned herein are the property of the trademark holder. We use them for identification purposes only. This is not an official publication.

Voyageur Press titles are also available at discounts in bulk quantity for industrial or sales-promotional use. For details write to Special Sales Manager at Quayside Publishing Group, 400 First Avenue North, Suite 400, Minneapolis, MN 55401 USA.

To find out more about our books, visit us online at www.voyageurpress.com.

Library of Congress Cataloging-in-Publication Data

Bowers, Steven, 1954-2007
  Driving horses : how to harness, align, and hitch your horse for work or play / Steve Bowers, Marlen Steward. — [2nd ed.]
      p. cm.
  Other title: How to harness, align, and hitch your horse for work or play
  Includes index.
  "First published as Farming with horses in 2006 by Voyageur Press, an imprint of MBI Publishing."
  ISBN 978-0-7603-4570-2 (softcover)
  1. Draft horses. 2. Draft horses—Equipment and supplies. 3. Agriculture. I. Steward, Marlen, 1933- II. Title. III. Title: How to harness, align, and hitch your horse for work or play.
  SF311.B67 2014
  636.1'5--dc23
                    2013026536

Editors: Amy Glaser, Elizabeth Noll, and Madeleine Vasaly
Design Manager: Cindy Samargia Laun
Design and Layout: Chris Fayers
Front cover and title page photos courtesy of Nathan Bowers

Printed in China
10 9 8 7 6 5 4 3 2 1

Contents

Acknowledgments

I couldn't have created this book without the lifelong and day-to-day support of my wonderful wife, Peggy. Her trust and faith in what I do for a living are a major part of what keeps me going. Our children, Katie and Nathan, are getting older now, which makes them more valuable as coaches of their father's writing projects. Marlen Steward, my partner on this project, continues to amaze me with his solid work ethic and insistence on presenting only the highest quality of illustrations. Midwest Leather, one of the largest and most customer-oriented harness makers, of Beckwourth, California, provided samples of the many different styles of harness shown in this book. Friends Larry Helburg, Gene Hilty, Dennis and Jean Kuehl, Jean Brandenburg, Don Faulkner, Les Barden, and Cherry Hill provided support and encouragement in various ways. Thank you.

Introduction
Hitched to a Purpose

Most modern-day people are accustomed to using power sources that can be characterized as cheap, fast, and easy. An obvious purpose of hitching up with real horsepower is to convert to a clean fuel from a clean source. Horses are designed to function very efficiently on grass, water, and salt. Other, more common power sources may seem cheap and easy, but they operate on fuels that are dangerous to be around, and their mining, refining, and transport is dangerous for the environment.

Many who use real horsepower have purposely tried to get away from making everything too easy. In a nation that is world famous for having a high rate of overweight young people and adults, the horse-powered Amish are a notable exception to that rule. When a Canadian study clipped pedometers onto these people, who are noticeably skinnier than the rest of us, they found that they move around three or four times more than those in the larger culture. If you own a tractor and you need to move a few bales out to the field to feed some calves, you walk over to the garage, sit on the seat, and start moving toward the hay stack. Doing the same job with horses requires feeding and caring for the horses in the morning before you do the job, and all the steps it takes to go get a horse or two, get them harnessed, and get them hitched to the hay wagon. Instead of stabbing into a big bale with a bucket fork on the front end loader like most tractor farmers, a horse farmer would pull into the barn and have to handle the bales one by one

or fork loose hay onto the hay wagon. By the time they get to the calves, the fast, cheap, and easy guy has burned about the same number of calories he was burning when he was sleeping. The horse-powered farmer or rancher is getting his calves fed while he is also getting a nice moderate workout.

If you were looking for a horse book that is like all of the rest, you might be a little disappointed with this one. There isn't page after page of pictures of horses being worked in the field on huge pieces of equipment. In fact, this book doesn't have any particular information on how to use even one piece of standard equipment. Before considering equipment, I think you should learn about driving and working horses from the horse's perspective. Your success or failure at driving horses is going to depend more on mastery of the subjects presented in this book than on how many pictures of working horses you've seen.

In this world, what many people are beginning to miss out on is a very important ingredient to happiness—having a great purpose. I believe that many people who are drawn to using horses are interested in this alternative way of doing things because they are getting a little sick of cheap, fast, and easy. An antidote to being caught up in all of the ills caused by our lifestyle is to quit doing things the cheap, fast, and easy way and switch over (in some way) to doing things with real horsepower. There are many compelling purposes to be accomplished by hitching up with real horsepower.

Chapter One

Functions of Harness

One of the difficulties of writing a book about draft- and driving-horse use is the fact that there are so many different types of harness out there. From the crudest of bamboo-and-rope neck yokes, as seen in developing countries, to incredibly ornate harness used by royalty, horses are worked in all kinds of different harness. One characteristic of all good harness is that no matter how it looks, it is designed to harness the power of the horse. To be a good harness, it usually needs to harness not only the obvious forward power of the horse, but also the turning, stopping, and reversing power of the animal. Rather than being distracted by the material of a harness or its ornamentation, a far better approach to learning about and using harness is to approach it with the viewpoint of function.

OVERALL PARTS

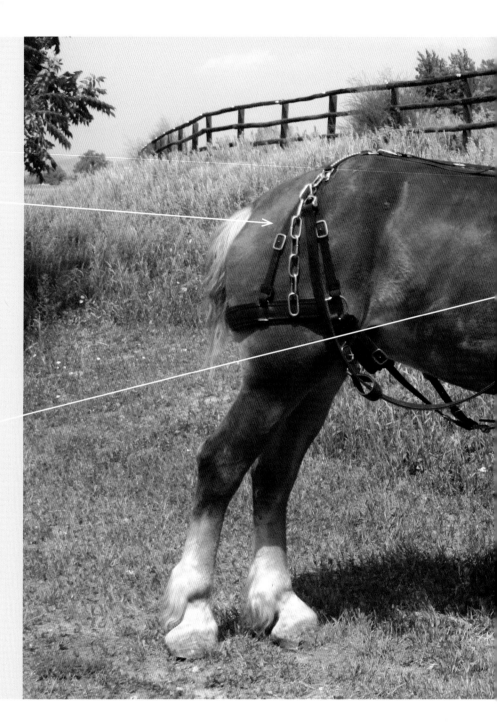

Hip Strap Assembly

Surcingle

I usually start teaching about harness by showing a draft-horse harness. The reason for starting with a draft harness is that all of the available directions of horsepower (forward, backward, right, left) are captured in ways that are obvious and sensible. In addition to draft-horse driving, there are three other major disciplines of driving, each with their own characteristic harness, commonly seen in America. These are pleasure driving, show ring driving, and harness racing. In this book, we will only be dealing with draft-horse driving and pleasure driving because these two disciplines use a harness that is most

Hames

Bridle

Lines

effective at harnessing the full potential of a horse's power. If you are going to be farming with horses, these are the harness types to use.

When you begin to get an understanding of how a harness is used on a horse, you will also begin to use the terms of each individual part of the harness.

Understand from the start that there are many different names for harness parts, no matter the discipline of driving. Even the pronunciation of the different names for the parts of the harness varies among horsemen. For years, I've wrestled with the proper pronunciation of the simple word *crupper*, which

BRIDLE PARTS

Side Check Rein

Gag Runner

Crown Piece

Brow Band

Blinder Stay

Blinder

Noseband

Cheek Piece

Bit Strap

Bit

Throat Latch

Lines

FRONT HARNESS PARTS

Back Strap

Ratchet

Collar

Hame Ball

Upper Hame Strap

Upper Hame Ring

Hame

Hame Bolt

Lower Hame Ring

Lower Hame Strap

Breast Strap

Coupler Snap

Pole Strap

Trace

REAR HARNESS PARTS

Hip Straps

Toggle Chain

Trace Carrier

Rein Up Strap

Back Strap Loop

Saddle

Breeching

Market Tug

Trace

Lazy Strap

Quarter Strap

Surcingle Keeper

Belly Band Billet

Pole Strap

Belly Band

Back Strap

▲ **Steering parts:** Line or rein, bridle, and bit. These parts of the harness control the movement (forward, back, stop, right, and left) of the horse as the driver dictates. The remaining harness parts use the movement of the horse's body to move the load.

is the piece that circles under the lower side of the tail on some harness styles. Some people say *crooper* (rhymes with *super*), and others look at you funny when you say it that way and correct you, "That's *crupper*" (rhymes with *supper*). Don't be intimidated by feeling like you need to know the name and the proper pronunciation for every part of harness shown on these pages. It is far more important to know how these parts function. If you are like most people, your vocabulary of harness parts will naturally expand the more you think about and use harness. Hopefully, the more you learn about the various names for parts of the harness, the less you will be inclined to correct others who say them differently.

VARIOUS TYPES OF HARNESS

To increase your knowledge of the various harness types available, pictures of some of the more common types are provided. Most of these harnesses are draft types of harness, because draft is by far the driving discipline with the most variety.

It's interesting to note that draft-type harness can be used on almost any breed of horse. Although typically seen on draft horse breeds, draft-type harness can also be appropriate to use on horses as small as Shetland ponies. Some of the distinguishing features of draft-horse harness are toggle chains on the ends of the traces; hames that adjust for length; and hames that have excess length (risers) above the top hame

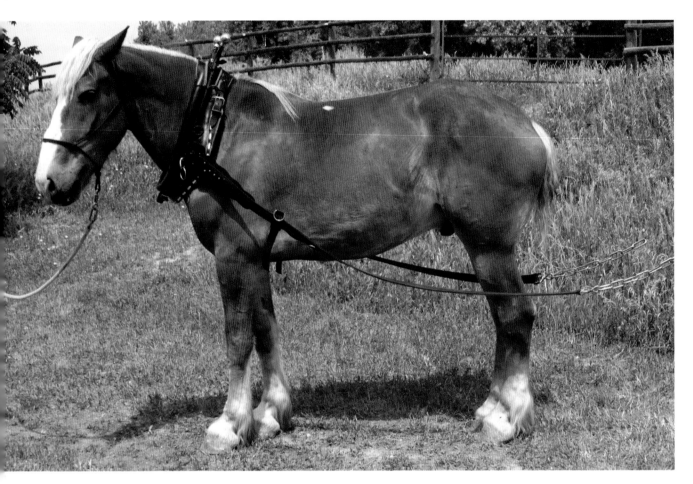

▲ Pushing and pulling parts. From the horse's viewpoint, he is pushing on the load. From the human's viewpoint, the horse appears to be pulling the load. For most jobs, the horse is primarily using the following parts of the harness: collar, hames and hame straps, and traces.

▶ Stopping and backing parts. These harness parts work along with features of the vehicles and/or implements to keep the load from running into the horse when the horse is stopped or backed up. It might help to look ahead to Chapter 4 to see how the harness and the vehicle work to keep the horse from being run over by the load.

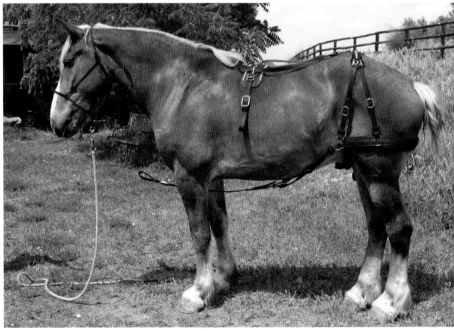

▲ **The turning parts.** For this type of harness, when used on a horse that is part of a team—two horses, side by side—most of the turning force that transfers the lateral movement of the horses to the lateral movement of the vehicle involves these parts. During turns that also involve holding the load back, the parts used for stopping and backing up (see photo to left) also come into play. When this harness is used on a single horse hitched to a vehicle in shafts, then the saddle, market tugs, bellyband billets, and bellyband are the active parts of the harness for turning. These parts of the harness that encircle the horse at his middle are collectively referred to as the surcingle.

strap, usually terminating in a ball ornamentation at the top.

In the same way, a pleasure-driving harness is usually seen on horses that are smaller than draft size. However, if a draft horse is to be hitched to a formal-looking vehicle, it can look nice if the draft horse is wearing a pleasure-driving harness. You can hitch your big draft horse to a light little vehicle like a road cart while your horse is wearing a heavy

duty–looking draft-type harness, but many people think it looks a lot nicer if you use a pleasure-type harness with a pleasure vehicle. If you are planning on showing your draft horse at a draft horse show, it's probably a good idea to go look at what everyone else is doing and copy their harness and vehicle selections as much as possible. Horse show judges are usually not famous for appreciating variation from the usual standards.

▶ **Box breeching or Western-style harness.** For comparison purposes, we're showing this harness again. According to my research, about 80 percent of the draft-type harness used are this style. This harness is made of a strong and durable material called bioplastic with a granite finish.

▶ **Hitch horse show-type harness.** This type of draft harness has many similarities to Western-style harness, but it is made with special features that are associated with showing and multiple-hitch, exhibition-type uses for horses. This type of harness uses high-quality hardware, leather, and stitching, and it is expected to see this type of harness maintained in perfectly clean and shiny condition. This harness is a little heavy to handle, so major sections are detachable for easier piece-by-piece harnessing and unharnessing. Another distinguishing feature of this harness is that it usually has a crupper. The crupper is the anchor point for the pressure needed to hold up or lift the horse's head with the overcheck. This harness is made of black-dyed leather. The shiny surfaces are made of patent leather.

◀ **New England D-ring harness.** This harness is an older type of Scandinavian origin that is usually only seen in a few places in the New England area. It has the amazing benefit of completely preventing any tongue pressure on the top of the horses' necks and transferring any downward pressure to the saddle, where the horses can more easily handle that pressure. Another great thing about this harness is how turning pressure bears on the horse at the D-ring, not at the overworked collar area. This is russet (undyed) leather with brass hardware. The hames are made of oak wood.

◀ **Yankee breeching harness.** In particularly steep parts of the country, you will sometimes see teamsters using Yankee breeching harness. When a horse is holding a load back on a steep incline, placing the breeching above the tail like this does a much better job of holding the load back without sweeping the rear legs forward, like a more conventional breeching placement. This is black leather with chrome hardware.

▶ **Side-backer harness.** This harness looks like a cross between New England D-ring and Western-style harness, but it is actually very similar to Western style. The difference is that the quarter straps are brought forward from the breeching, along the side of the horse, instead of down under his belly. One of the main reasons for this placement of the side strap is that it keeps the braking system of the harness up and out of the dripping sweat area of the horse's body. Day after day of sweating up the quarter straps and pole strap on a Western-style harness makes the leather on these parts begin to rot, which is something that doesn't happen with side-backer harness. This and the New England D-ring harness feature jockey yokes at the front of the harness instead of the usual coupler snaps. This is made out of bioplastic and has a shiny finish.

▶ **Pleasure-driving single harness.** This is a harness from a different discipline of driving than the harnesses shown above, but it has many similarities. Like draft-type harness, this harness is also made to maximize the effectiveness of equine effort. This harness has a breast collar, but many times a pleasure-driving harness will be seen with a collar and hames if the load is heavy. It's easy to notice the more elaborate, padded saddle. Better-quality single harnesses feature saddles with wood, fiberglass, or spring steel trees to help distribute the weight of the shafts over a larger area. This harness, like the others in this section, features a breeching to hold the load away from the horse. This harness is made of leather with brass hardware.

▲ **This is an "out of adjustment" harness.** Too long of a collar puts the point of draft too low and is a major cause of collar sores. Straps that are too long in the back put the braking parts of the harness too far to the rear, making the horse look long and low. Breeching pointing below the center of gravity of the horse makes the horse ineffective at holding back heavy loads. Harness adjustment affects the comfort level of the working horse and it also affects the appearance.

HARNESS FIT CONSIDERATIONS

One of the most useful things to know about harness fit considerations is that horses are sensitive to their harness fit. For the harness to feel good to your horse, it's important to remember that a horse's center of gravity needs to be balanced with the load forces. It's exactly the same as a human pushing a lawn mower: if a person is going to mow a large area or has to push very hard, it's important to fuss with the handlebar height until it's lined up with the human center of gravity, which is about at the belly button. If you've ever had to push a mower that had the handlebar set too high or too low for your height, you can begin to understand the frustrating feeling a horse will have if the forces he's trying to exert aren't lined up with his center of gravity.

My favorite way of explaining center of gravity on a horse may seem a little crude, but it's effective. Imagine a horse standing at attention, ready to work, with his head held neither too high nor too low. Now imagine freezing the horse into a solid statue with a painless procedure. Further imagine piercing the horse through with a steel skewer from the side so that the skewer goes exactly through the horse's center of gravity. If you found the exact right spot, that steel skewer could be held horizontal while lifted vertically about 6 feet above the ground. In theory, if you went over to the horse's tail, which is now hanging a little above your head, and pulled a hair out of the tail, this frozen horse would smoothly begin to spin around the skewer with no further tail pulling needed. That place where the skewer would go through to achieve such a feat is called the center of gravity of the horse. On a live horse wearing a dressage saddle, the center of gravity is under the stirrup leather about one-third of the way up from the ventral line of the horse. On a horse wearing a well-fit harness, you can find the center of gravity by locating the intersection of the surcingle (the harness parts that encircle the horse's trunk at the saddle) and the trace.

When the trace pressure goes through the center of gravity on a horse, the horse's strength on the trace is maximized. If the horse is improperly harnessed or hitched so the trace is too low or too high relative to the center of gravity, then the horse will feel like he can't effectively apply himself. On light vehicles, this principle of having the trace cross over the center of gravity of the horse is not nearly so critical as it is if the horse is expected to exert himself on a heavy load.

Likewise, the pressure from the breeching should also point toward the center of gravity on the horse. If the breeching is too low, it points under the center of gravity and makes it impossible for the horse to fully use his power. Another problem with a breeching that is too low is that the forward pressure can sweep the horse's rear feet out from under him.

When a horse is harnessed, it greatly increases the visual appeal of the horse if the ring at the top of the hip strap assembly is positioned a bit forward of the high point of the horse's rump. Poor harnessing can make a short-backed horse look strung out like an alligator. With skillful harness adjustment, even a long-backed horse can look well-built and short-coupled.

HARNESSING A DRAFT HORSE

There's more than one way to harness a draft horse. For this book, I will focus on how I like to see harnessing done on a horse. I know an Amish horse trader who likes to show you how well broke his teams are by

▲ **Complete harness on hook.** A complete "side" of draft horse harness hung up after previous use. Notice that the collar is hung upside down for strength on the hook, with the surface that was toward the horse exposed to the air for rapid drying. The bridle is also hung upside down by the bit instead of the crown piece. These heavy draft horse bridles with blinds tend to look very droopy if they are hung by the crown, as other kinds of bridles are hung. The main part of the harness is hung by the top hame strap at the front of the harness and by one side of the hip strap assembly at the rear.

◀ **Collar going on.** The easiest way to put a collar onto a horse is to simply ease it on over the horse's head. It helps if you spread the collar between your hands and knee before doing this. If your horse has never had a collar put on or taken off this way, be sure to have a lead rope on the horse and don't be tied up in case the horse needs to move his feet. Some teamsters make it a practice to turn the collar upside down as it passes over the horse's eyes, then they rotate the collar right side up at the throat latch. There are some horses whose conformation makes their heads wider than their necks, and the collar will physically not go over the horse's head like this.

◀ **Collar being opened.** To open a collar and put it on, undo the fastening at the top of the collar and be careful to support both sides of the collar so they don't fall apart when opened. Collars are stuffed with rye straw to give them their shape. Opening a collar too far breaks the rye straw at the throat of the collar, causing a loss of shape that the collar needs to perform as intended. The collar can then be buckled together at the narrowest part of the horse's neck and slid back into position. There's more about collar fit considerations in Chapter 2.

▲ **Harness the human.** Before you can harness a horse, you need to get the harness onto yourself properly. Start by removing the top of the hip strap assembly from the hook and placing it onto your right shoulder.

▲ **Hold the hames.** Where you hold the hames for harnessing can make a huge improvement in how easy it is for you to get the harness onto a tall horse. After reaching under the saddle with the right forearm so the saddle is supported there, grasp the right hame midway down its length. At the same time, grasp the left hame at the same place. Harnesses are usually hung on a high hook to make sure that they are not dragging on the barn floor when hung up.

literally throwing the harness onto the horses from about 4 feet away. After a few years of watching this performance and seeing its result the next time the horses were harnessed, I finally got it figured out that I should tell him to not do it that way. Now, when this guy harnesses a horse that I might be interested in buying, I always say, "Elmer, could you please just set the harness on nicely? I might end up buying this horse and I don't want to spend the next three months having the horse duck when I lift the harness!"

Many folks new to harnessing, especially short people, figure that they'll need to pass the harnessing job off to someone taller in the family, especially if the horses are 18 hands tall or more. If you harness in the way shown here, being careful to get hold of the

hames, as shown in these pictures, it is possible for short people to harness big horses.

Many people who harness bigger horses make things more difficult for themselves when they try to pass things over the horse's back that don't need to go over, or they get in too big of a hurry and start pushing on parts that ought to be left alone. By following the process outlined in these pictures, you can see the details that will make harnessing an easy experience for you and your horses.

▲ Putting the right hame into place. When you begin to place the harness onto the horse's back, try to focus simply on getting the right hame up over the horse's shoulder. This means that your left hand needs to come up a little to support the activity, but not much.

▲ How to approach. Don't walk up beside a horse while carrying a harness until you are absolutely certain that the horse sees you coming and appears to be alright with you approaching. If the horse doesn't see you coming or if the horse appears alarmed at your approach, don't touch the horse. When it seems okay with the horse, walk quietly to the horse's shoulder without doing any hame lifting. If your normal approach to harnessing involves walking up along the horse's side while simultaneously lifting the hames above your head, you will soon meet a horse that thinks it is scary, which might get you kicked. It calms a horse to pause for a bit at the shoulder and tell him how much you like him before proceeding. When walking up to a horse carrying a harness, your eyes should be on the horse's ears. The horse's ear follows his eye movement, and the ear is quite a bit easier to see. Many teamsters give a sharp "Whoa" command as they walk up to a horse with a harness, but I try to not do that because giving a command for something that is already being exhibited numbs the command response. If the horse doesn't see me coming, I might quietly say his name or clear my throat and make sure he sees me before proceeding.

▲ Chase the trace. Now that you have placed the right hame up there as far as you can reach, it is time to release your hold on that right hame and move your right hand over onto the right-side trace. With your right hand on the trace, use the stiffness of the trace material to "chase" the trace over onto the other side of the horse. By repeatedly grasping and pushing the trace over onto the other side of the horse, the harness ends up centered on the horse's back without a huge amount of effort.

▶ **Leave the back straps loose.**
Don't be in a big hurry to get the entire harness positioned perfectly onto the horse right away. If you did that, it would tighten the back straps and make it difficult to get the hames positioned on the collar. You can relax now because the heavy lifting is over and it's time to think about fastening the harness onto the horse.

▶ **Fasten the hames.** Starting at the front of the horse when harnessing and beginning at the rear when unharnessing are great habits to get into. That way, if a green horse ever starts jumping around when the harness is only partially attached, it is going to be fastened at the hames, which makes it likely that the harness will stay with him. When fastening the hames, you want them to fall into place in the groove of the collar and then use some muscle to fasten the lower hame strap, pulling it into the buckle with great firmness. If you are going to fasten your hames from the left side, hang your lower hame strap from the right hame, as seen in these pictures so the teamster can stand on the left side of the horse and comfortably pull toward himself for tightening the buckle.

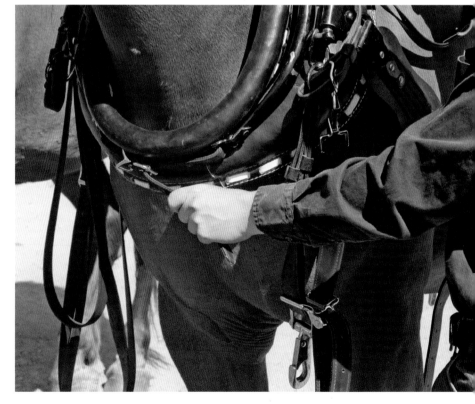

▶ **Fasten the breast strap.** Some teamsters have their breast strap, coupler snap, and pole strap hung from the right hame, so they throw all of this material over to the right side of the horse when harnessing. It's much easier when harnessing from the left side of the horse to anchor your breast strap to the left hame so all these parts stay on the left side of the horse and don't have to be pushed to the other side. Some teamsters have snaps on both ends of the breast strap, but those breast strap snaps are a weak point and could cause trouble if a horse happened to get his bit ring caught in one. The breast strap is designed to allow the coupler snap to slide throughout its length, so be sure the breast strap isn't twisted as it crosses from one lower hame ring to the other.

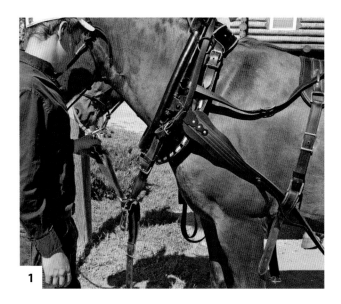

▲ **Put the bellyband through the pole strap loop.** Better-quality Western-style harness will have a loop in the end of the pole strap for the bellyband to pass through as a safety mechanism (you can read more about it in Chapter 8). If your harness doesn't have a loop in the end of the pole strap, put the pole strap above the bellyband before fastening it.

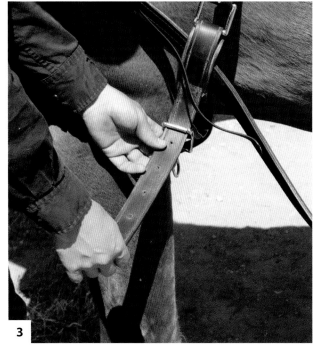

▲ **Fasten the bellyband.** On draft and pleasure-driving harness, it's not usually necessary to extremely tighten the bellyband like a saddle is tightened on a riding horse. The bellyband can be left hanging rather loose. It serves no real purpose to have it tight. The bellyband and the bellyband billets prevent any trace lifting due to load orientation, such as when the horses go into a ditch ahead of the wagon, which keeps the collar and hames from lifting up when the traces are lifted.

▶ **Pull the hip strap assembly into place.** Now you can pull the back end of the harness into position over the horse's rump. The place for the tail is out over the top of the breeching, not under it. When you pull the tail out from under the breeching, be sure to sweep the tail bone and tail hairs to the side to get over the breeching. Tail bones aren't made to flex vertically, so if you stand at the rear and pull the tail straight toward you and over the breeching, you'll make an unhappy horse. On horses you don't absolutely trust, it is good practice to keep your knees pointed in the same direction as the rear of the horse in case the horse kicks. Getting kicked in the back of the knees is a lot better than getting kicked on the front side, orthopedically speaking.

▶ **Fasten the quarter straps.** Quarter straps have definite up and down sides to their snaps. The smooth side of the snap is the side that doesn't open. The snap's smooth side is snapped into the ring in the end of the pole strap so that it is up toward the horse. If the opening side of the snap is pointed toward the horse, he will object because that side of the snap hurts when it is laid into the animal's sternum.

▲ **Check the final fit.** One final step to harnessing is to be sure the quarter straps and pole strap are adjusted to put the coupler snap in a particular spot. When the coupler snap is pulled forward, its forward limit of travel should be on an imaginary line defined by the hame when viewed from the side. To test for this, you first have to be sure your horse is neither standing stretched out nor with his feet bunched together. With the feet squarely under his body, you can check for final fit.

◀ **Put on the bridle.** If you go to look at a driving horse and they bridle him with a blind bridle first, that should be a warning to you that the horse probably isn't safe to harness unless he can't see you. We usually only bridle after all the rest of the harness is on the horse and we are ready to head out. If your horse is going to wear a blind bridle, you need to be aware that he can't see his surroundings very well once the bridle is on. It may be a better idea to bridle outside the barn with some horses, especially if the doors are narrow or the alleyway is cluttered with equipment or feed. We start bridling by getting the bridle in the "ready" position. The side check rein is placed over the horse's ears first. The bit is being carried into position with the right hand, which is holding the crown piece of the bridle. With the bridle in this position, the head can be controlled fairly well with the elbow against the cheek to push the head away as needed. Lateral pressure with the right hand can also be used to draw the horse's nose closer, if needed.

◀ **Get the bit in the mouth.** By lowering the right hand a little, the left hand can direct the bit toward the horse's lips. Use just enough pressure with the right hand lifting to keep the bit positioned lightly against the lips, while the left hand begins to open the horse's mouth. There is an interdental space that doesn't have any teeth between the incisors and the molars on young horses. On older (past 4½ years) geldings and stallions, that space can be a little thorny with canine teeth, so you have to be careful on some horses, but right there at the corner of the mouth is the safest spot to insert the thumb of your left hand while it also holds the bit. When your thumb insertion works properly, it will cause the horse's teeth to come apart far enough for you to get the bit past his incisors by lifting with the right hand. Do all of these activities slow and easy because most horses have an aversion to rough handling around the mouth, and most resistance to bridling is caused by poor handling and rushing the job.

▶ **Switch hands.** Now you need to get the top of the bridle into your left hand so that you are prepared to get the crown of the bridle over the horse's ears.

▶ **Do the right ear first.** Many horsemen put the bridle over the left or near ear first, but that leaves them with the problem of having to get the bridle over the right ear of the horse, which is further away. If you do it that way, when you reach to get that far ear you have the additional difficulty of having to get the bridle over that ear while the headstall has the slack taken out of it by the left ear being bridled already. It's a mess that way. The smart way to bridle a horse is to hold the headstall forward with your left hand so the right ear can be eased forward with the right hand. With the right ear laid down forward and out of the way, it's pretty easy to slide the headstall into position over your right hand.

▼ **Do the left ear.** Once that difficult far ear is holding up the bridle, you only have the much easier left ear to deal with. Again, fold the ear forward with a flat hand, bringing the headstall forward enough that it clears the ear without touching it, and bring the headstall back over your hand into position behind the ear. Many teamsters shove the top of the bridle back over both ears at once and then pluck the ears and the foretop out from under the bridle with lots of folding and crimping of the ears. Not all horses will accept being treated that way.

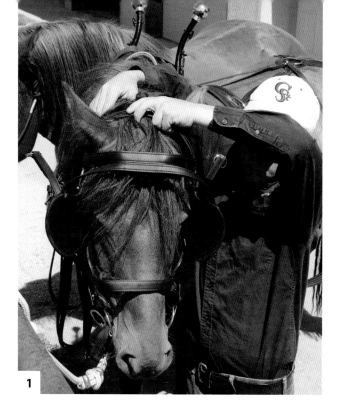

1

2

▼ **Secure the throat latch.** If your horse is in a blind bridle, you definitely don't want anything to happen where the bridle comes off an ear, which would expose an eye and almost certainly cause a bad wreck. Throat latches on driving bridles should be fastened tight enough to assure that the bridle won't rub off if the horses get to itching on each other or your hitching post. Fastening a throat latch too tight might cut off a horse's wind while working, so try to find a balance between too tight and too loose. You can test the effectiveness of the throat latch by fastening it and then using your hand to try to remove the bridle with the throat latch fastened. When you get the right tension, you won't be able to pull the bridle off, but it won't be too tight on the throat.

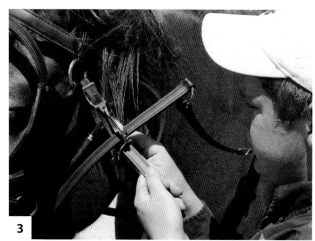

3

▶ **Fasten side check rein to rein up strap.** Be sure to pick up the coupling line, which is lying near the top hame strap, and go under it with the side check rein. Side check reins are on working horse bridles mainly to prevent untimely grazing. They also seem to help prevent two-footed kicking while hitched. Some horsemen who are working horses try to adjust the side check rein and the rein-up strap so the horse isn't being interfered with while doing his work. For draft horse showing, show-ring driving, and harness racing, horses are often driven in an overcheck (rein that goes up the front of the face and between the ears, then down the top of the neck), which is a much more effective way of lifting heads to an artificial height. In sports where heads are lifted with an overcheck, you also need to use a crupper under the tail as an anchor point for the head lifting pressure.

◀ **Pull the lines down.** Before you hitch to anything with your team, you need to unfasten your team lines from their storage position on the outside hame of each horse. The lines are already threaded through the upper hame rings on each horse's shoulder so all that is needed is to stand the team side by side and begin to fasten lines to bits.

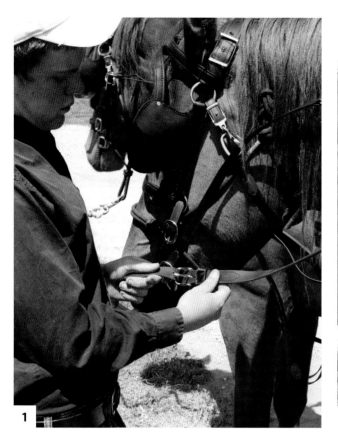

▲ **Draft lines to bits.** The draft lines that are to the sides of the team fasten into that side of the bit. The left horse has the left draft line fastened into the left side of his bit. The same arrangement works for the right horse's right draft line.

▲ **Fasten the coupling lines.** The coupling lines need to cross to the other horse in front of the middle hame rings. I've seen teams driven with the coupling lines crossed behind the hame rings, but that doesn't work so well. Now your team is fully harnessed and ready to be ground driven or hitched.

UNHARNESSING A TEAM OF DRAFTS

Many folks would think unharnessing a team of horses might take a lot of time, but I know from firsthand experience that a big team of drafts can be unharnessed in a bit less than 2 minutes without much effort. We used to have unharnessing races after the sleigh rides each night at Nordic Sleigh Rides in Breckenridge, Colorado. Our competitive group of drivers would unharness the three teams one at a time, tied to the fence right in front of the harness room. We kept track of the best times by writing them on the drywall in pencil inside the harness room. My friend Pete, the nimble-fingered mountain climber, usually won; his best time ever was 1 minute, 40 seconds. The team started the race fully harnessed and collared, wearing bridles and lines still drawn to the rear for driving. At the finish, both horses were only wearing their halters and the harness was fully hung up in the harness room.

For well-trained horses that are paired up as a team, unharnessing usually starts by repositioning the bit ends of the lines from the bits to the lower hame rings. Some teamsters completely remove the lines when unharnessing, but the way shown here saves a lot of time when unharnessing and harnessing. Leaving the lines on and making sure they are always left attached to the outside hame provides a handy way of making sure that the harnesses go onto the correct horses when harnessing.

▶ **Pull the slack out.** Any excess slack is pulled out of the forward part of the lines as the draft and coupling line is drawn forward (from back to front) between the hames. Doing it this way is very important because this step keeps the lines from getting under the hames the next time the horse is harnessed.

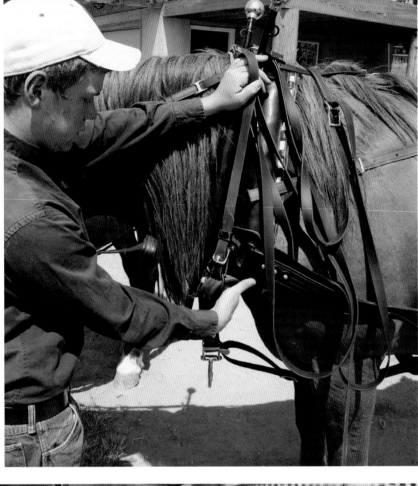

▼ **Fasten the lines to the lower hame rings.** After removing the ends of the lines from the bits, they are re-fastened to the lower hame rings of the harness, returning the lines to their starting position on each horse.

▲ Folding again. Drop down the fold you just made until it hangs level with the end of the lines and create a second fold in your hand.

▲ Make the first fold. Find the middle between the coupling line buckle and the driver's end of the line. Temporarily mark this midpoint by folding the line.

▶ Set the lines back of the hames. You are now ready to hang the lines on the hame. First make sure that the unfolded part of the lines, near the hames, are drawn to the rear and behind the hames. This move also helps prevent lines from getting under the hames when harnessing next time.

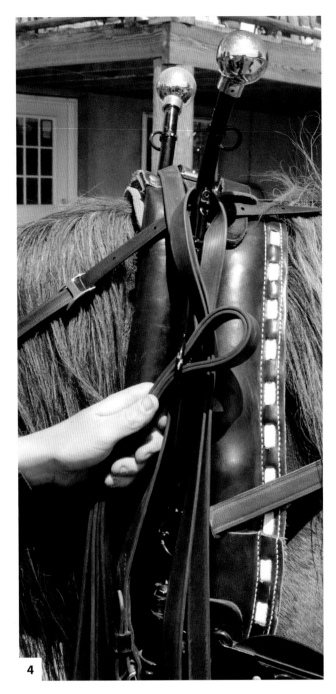

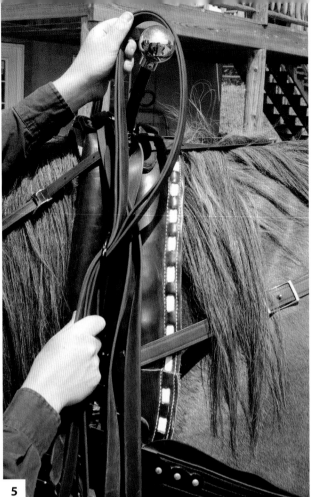

▶ **Over the hame top.**

▼ **Lines fully stored.**

▲ **The fold is looped over the top of the hame.** Then drawn down until it rests securely against the upper hame strap. There are several other methods of hanging up your lines, but I like this one the best. By always putting your lines up on the outside hame instead of the hame that is between the horses, it is easy to tell which harness goes on which animal when harnessing.

◄ **Take the bridle off.** As properly done when unbridling a saddle horse, stand to the left side of the horse's neck and use the left hand to carefully draw the top of the bridle off the horse's head. Start by placing any side check rein that might be present near the back of the horse's ears so the rein is removed with the bridle in one smooth move. The right hand and forearm are used to ward off any head rubbing efforts by the horse. Don't drop the bridle at this point. Wait for the horse to open his mouth and release his hold on the bit before lowering your left hand.

▶ **Hang the bridle by the bit.** If you lived in a zero-gravity part of the universe, you wouldn't have to hang your bridle this way. In all other locations, it makes sense to hang your bridle so that the effects of gravity rejuvenate the look of your bridle. Droopy bridles greatly depreciate the look of both your horse and harness.

1

▲ **Quarter straps off.** It's a good idea to get into habits that will serve you well later. If you always start here and work your way forward, then the last thing attached when unharnessing is the hames. Having the hames attached until the last moment of harness wearing means that someday, when you are unharnessing a colt and he starts jumping around with your harness on, it will have a good chance of staying where it belongs.

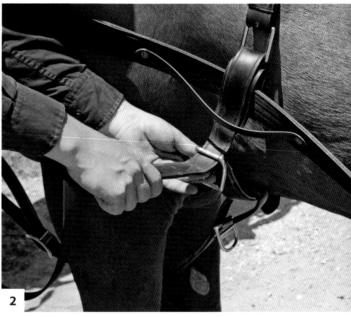

2

▲ **Bellyband and pole strap unfastened.** Moving forward, the next thing to undo is the bellyband. If there is a loop in the end of the pole strap, the bellyband is backed out of it.

3

◀ **Breast strap.** You won't get your harness off unless you undo the breast strap and bring it back to the near side of the horse. It is snapped into the left lower hame ring. If you are driving a team, both horses are usually harnessed and unharnessed from the left or near side. This is a carryover from the rider's custom of mounting and dismounting from the left side of a horse. It all got started with the fact that most warriors were right-handed and had the long sword in its scabbard on the left hip for easy drawing.

▶ **Bottom hame strap unfastened.** Complete the unfastening of the harness by undoing the bottom hame strap. Now the harness is loosely lying over the horse and ready to be moved to the harness hook. Many teamsters step to the rear of the horse and begin to lift the breeching onto their shoulder as they take the harness off from back to front. This method is not a good idea, mainly because it's difficult. It also puts the manure and urine-coated breeching onto your clean shoulder, and it involves movement that pushes against the horse's natural hair lay.

4

5

6

▲ **Top hame strap onto left forearm.** Use your right hand to grasp the right hame across the horse's neck and begin to pull that hame toward you. After a bit of pulling, you'll create a gap under the top hame strap where you can insert your left forearm and begin to carry the harness to the rear.

▲ **Saddle onto left forearm.** Continuing the movement toward the rear with the harness, the next part that is carried is the saddle, which is also placed onto the left forearm.

7

▲ **Take the hip strap assembly in your right hand.**
As the teamster moves to the rear, the next major part
to be carried is the hip strap assembly. Grasp the off-side
of the hip strap assembly with your right hand and begin
to carry it toward you as you step to the rear, sliding the
entire harness off the horse.

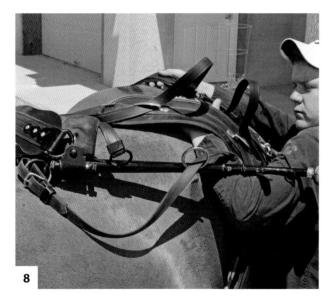

8

▲ **Walk it off the horse.** If the horses are wearing summer
short coats and have been sweating, it'll take a little effort
to drag the harness to the rear. Well-trained horses seem
to appreciate the feel of a harness that is removed this way
as they stand quietly. The movement of everything agrees
with their hair growth like a good, all-over brushing.

9

▲ **Harness yourself.** The number one reason some
teamsters unharness from rear to front is because they
don't know how to do this part. In order to get the harness
properly onto a harness hook, you have to know how to
harness yourself. Start by getting the hip strap assembly
onto your right shoulder, then reach under the saddle and
get the right hame in your right hand. The left hame goes
in your left hand without going under the saddle first.
You are now ready to carry your harness with a degree of
comfort and without looking too scary to any humans or
horses in your path.

▲ **Hang up the harness.** Hang the harness onto a wall-mounted hook as if the harness were hanging on a short-backed horse that is facing the wall. This allows for easy inspection while the harness is hanging up so you can tell what goes where.

▲ **Hang up the hip strap assembly.** Don't hang the saddle on the hook. If you hang the saddle on the hook, it unnecessarily supports the traces near their middle and makes them stick out into the alleyway of your barn or tack room. Hang the off side of the hip strap assembly onto the hook. This way of hanging allows for easy harness inspection and keeps it ready for easy harnessing.

◄ **Take the collar off.** It's nice if your trained horse will allow you to slip the collar off over his head, but some horses' heads are just too big to allow that. Collars can usually be unfastened at the top. The use of collar pads complicates the collaring and uncollaring processes. If your horse is wearing a full collar pad, the usual method is to unfasten the pad clips on the left side of the collar, then unfasten the collar top, and slide the collar up the horse's neck for removal.

ALTERNATIVE WAYS OF STEERING A TEAM
Single-Line Leader

This is the way my father drove singles, teams, and multiple hitches when he was a boy on a farm in Maryland. It is also what is sometimes referred to as "jerk-line" steering. The single-line leader or jerk-line horse is the only one in the entire hitch that is actually being driven by the driver's hand, with only one line coming back to the hand. The horse is taught to go right, left, stop, and back by signals from the single line. An ordinary team line can be used for this by adjusting the line so that draft and coupling lines are of equal length.

▲ **Path of single lines.** If you are hitching your horse as a single to an implement, vehicle, or load, the next step in harnessing is to attach the reins to the bit. On most harnesses that have hames, the reins are passed through the upper hame rings on either side of the horse's shoulder, allowing for driver-to-horse (and horse-to-driver) communication about direction and speed.

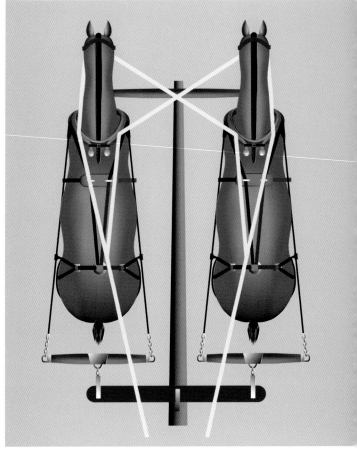

▲ **Path of team lines.** Pairs of horses that are hitched side by side and wearing draft-type harness are called "teams." The same two horses wearing pleasure-driving pair harness is called a pair. There are several different ways to rein up pairs of horses, but by far the most common is the one shown here, called team or coupling lines. The line that runs in a continuous length from the lateral side of the bit, through the upper hame ring on that shoulder, and back to the hand of the driver is called the draft line. In draft-horse driving, those leather ribbons you steer with are called lines. In all other disciplines of driving, those controlling devices are called reins. The line that buckles into the draft line, goes through the upper hame ring, and crosses over to the medial side of the bit on the other horse is called the coupling line. It is also known by many other names, such as crossover line, check line, stub line, and brace line. Pulling the right draft line drags both heads to the right. Pulling the left line brings both heads equally to the left. Pulling back on both lines stops the horses.

▶ ▼ How to turn with single-line.
To turn the horse to the left, the line is flipped over to the right side of the horse's body and pulled to the rear, which turns the head to the left. Flipping the line to the left side of the horse's body before pulling turns the horse to the right.

▶ To stop and back up, the line is held straight down the middle of the back. Well-trained horses eventually respond to a combination of voice commands and steady or jerky pulls of the line. "Gee" with a jerky pull on the line means "go right," and "Haw" with a smooth, on-and-off pull means "go left."

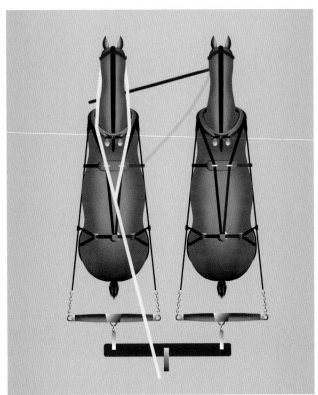

Jockey Stick

Many horses used in farm-type multiple hitches are controlled by a jockey stick. If you were driving a team and using a jockey stick, it would look like the picture below. The well-trained horse can be driven either by a single line or by a pair of single lines. The jockey stick goes from a lower hame ring on the anchor horse out to the outside bit ring of the stick horse. The inside bit ring of the stick horse has a strap or rope attached that goes back to the market tug/trace junction of the driven horse. Jockey sticks are also used on teams wearing traditional coupling lines to provide stiff arm resistance in cases where one horse wants to get too close to another.

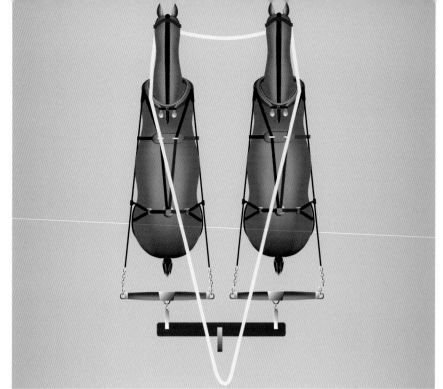

European Loop Rein

In some countries, horses are driven this way for field work. The steering rein is simply a big loop, usually made of rope. With this way of driving, it doesn't really matter where the driver stands. With more traditional coupling lines, it is important that the driver positions himself in the middle between the horses.

Achenbach Reins

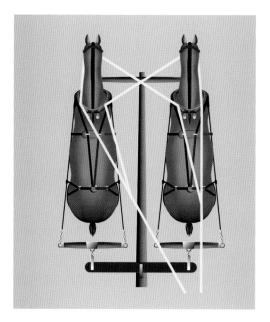

This is a particular type of coupling reins for teams and pairs that allows the driver to sit in a position above and behind the right-hand (off) horse when driving. Standard reins and lines are made for the driver to sit in the middle between the horses. Any movement from that position affects the characteristics of the lines and can cause warped results. When given a choice, drivers usually sit on the right side of the seat (as viewed from the rear of the vehicle). If you are carrying a whip, it is in the right hand and passes across in front of the driver's body. The excess length should be sticking out to the left side. If the driver is sitting on the left side, as if he is looking for the steering wheel, the whip would be carried too far to the left; if the whip is too far to the left, it will get in the way of traffic and stationary objects. Achenbach reins are made to have a definite left and right rein, so they're punched differently than normal team and pair reins. The rein with the buckle at the driver's end is usually the right rein.

Postilion

Artillery horses were mostly driven this way and it seems to be a popular way of handling horses for royal processions. The driver is also the rider and sits on the near horse to control that horse and the one beside. Both horses are steered by individual riding reins looped around the horses' necks. You can practice this way of steering with a bit of safety by not being hitched at first. The strap seen joining the horses together is simply a breeching tie. Postilion can be a nice way to get your team to the field and back when the implement is left in the field and it is a long walk for the teamster. Be sure your horses are comfortable with the implement you are pulling before driving postilion, because this would be a bad position to be in with poorly trained animals. Falling off of the horse would only be about half the wreck.

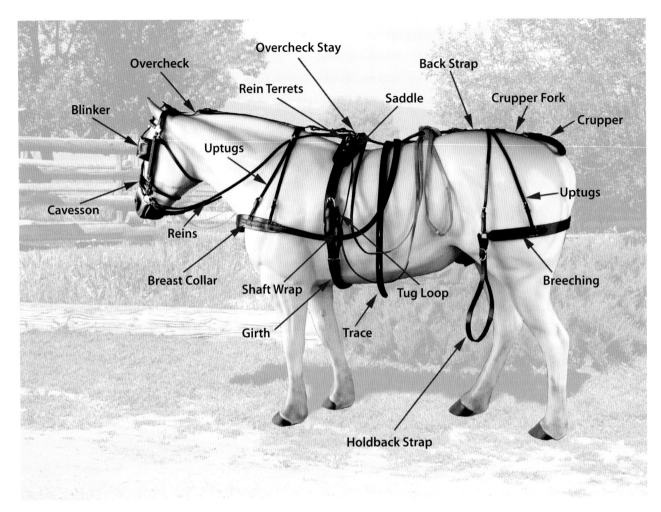

Names of pleasure-driving single-harness parts.

THE PLEASURE-DRIVING HARNESS

Once you are familiar with the look of draft-type harness, it is easy to see some similarities in the function of pleasure-driving harness. Pleasure-driving harnesses, like draft-horse harnesses, are designed to maximize the functional ability of the horse. A pleasure harness is built a bit lighter than draft harness, but that is okay because the usual purpose for this type of harness is for speedy and light travel rather than field work. If harness makers leaned too heavily toward maximizing the horse-to-harness surface area in order to make the pulling, stopping, and turning easier on the horse, they also would cause the horse to be far too covered up to cool properly on a hot day.

The cooling issue argues for the harness to be as spare as possible, especially when the loads are lighter.

This pleasure-driving single harness is used with a vehicle that has a shaft on each side of the horse. The shafts hold up the front of the vehicle when hitched to a cart. The shafts also provide for steering and stopping of the vehicle, as they follow the movement of the horse.

A pleasure-driving single harness is equipped either with a breast strap or with a collar and hames for pulling the load. Unlike on draft-type harness, the pulling apparatus is separate from all the rest of the harness, allowing for different options. The breast

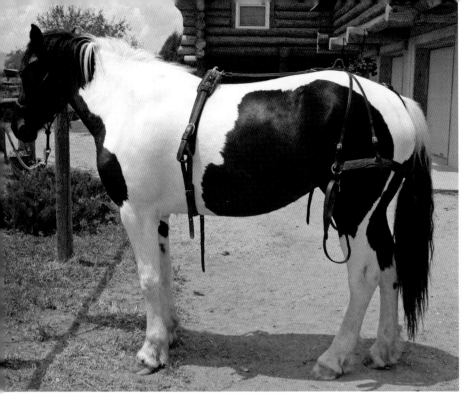

◄ **Put on main part of harness.**
It seems to work easier if you put the main part of the harness on first. The surcingle is buckled around the horse from the left side of the girth, leaving the girth somewhat loose to allow for side hill use of the horse and vehicle. Surcingle surrounds the horse at the horse's center of gravity. The saddle is placed at the base of the withers. The back strap is drawn rearward and the crupper is placed under the horse's tail by unbuckling the crupper from the crupper fork as it is positioned. When adjusted properly, the back strap won't tilt the saddle to the rear too much or show much slack.

▶ **Breast strap and traces.** For lighter-weight vehicles and loads, a breast strap or breast collar can be used. It is adjusted by lengthening or shortening the neck strap so the top edge of the breast strap is up near the trachea, but doesn't push into it at the horse's front.

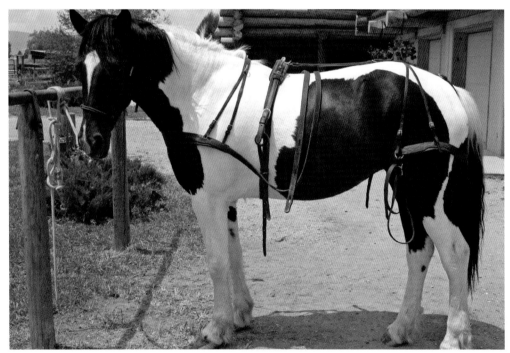

strap has the advantage of being able to fit a wide range of horse sizes, but it is not as good of a surface for effective pulling as the collar provides. One of the distinguishing characteristics of a pleasure-driving harness is the all-leather traces with slots in the end for attaching to the vehicle. A draft-type harness has toggle chains on the ends of the traces. Another distinguishing characteristic of a pleasure-driving harness is the appearance of the hames. Pleasure-driving hames don't have the risers above the top hame strap, and they don't have the hame-adjusting ratchet seen on draft hames.

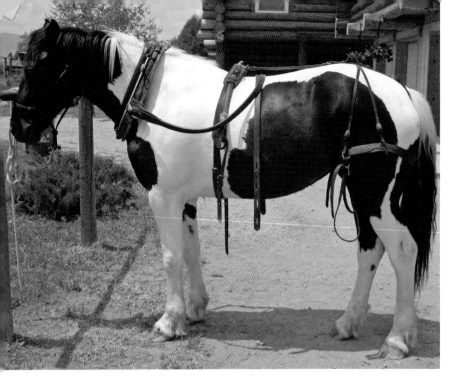

◄ **Collar and hames.** Substituting a pleasure-driving collar and hames for pulling apparatus makes the horse more comfortable when pulling heavier vehicles. Common collar sizes for saddle horse size horses are 18, 19, and 20 inches. Since pleasure-driving hames are not adjustable like draft-type hames, the hame length usually has to match the collar measurement (18-inch collar = 18-inch hame). It would be easier to make do with a hame that is a size too small than one that is a size too large for the collar because letting a top hame strap out can make up for an inch. A hame that is too long for the collar won't fit the hame groove.

▶ **Bridle on, with side check fastened.** Rather than using an overcheck to lift a horse's head, the side check is intended simply to prevent untimely grazing. It also cuts down a horse's kicking ability somewhat because they have a hard time doing double-barreled kicking with both hind feet when their head is not between their knees. The sight-restricting devices are called blinkers with this type of harness. On draft horse harness, they are called blinders.

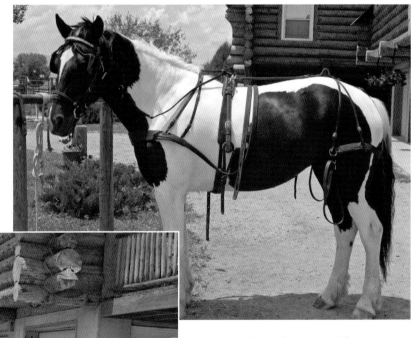

◄ **Reins through terrets.** Pleasure-driving enthusiasts call these things reins. On a draft-horse harness, most people call them lines, although I often slip up and call them reins.

Hanging Reins in Ready Position

This is a neat way to hang up your excess rein length on a pleasure-driving harness. When the driver is ready to get on the vehicle, the end of the reins are pulled and they smoothly release, returning the reins to a fully functional position.

▲ Pass the fold under the back strap.

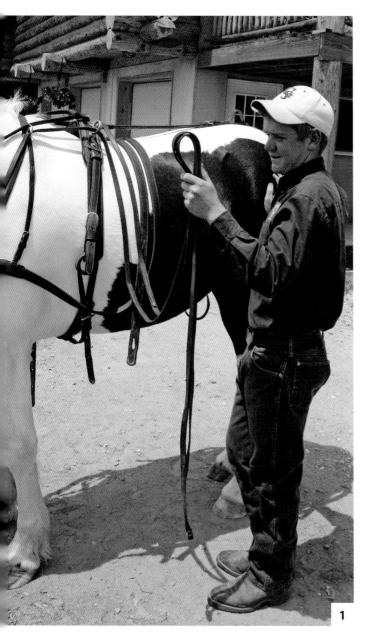

▲ Find the middle between the rein turrets and the end of the reins.

▲ Make another fold toward the end of the reins. About a foot from the rear part passing under the back strap, make another fold in the reins.

4

▲ Stick this fold through the first fold. Pull on the side going toward the bit to tighten this quick release set-up.

5

▲ The reins are ready to use. When you are ready, a pull on the using end of the reins neatly releases them into the hand of the whip. Pleasure drivers are usually called whips because they drive with a whip in one hand. Draft-horse drivers of singles through multiple hitch are called teamsters.

Why the Girth is Left Loose

If you grew up learning horsemanship only as related to the ridden horse, you might have a hard time understanding why a girth, such as on single pleasure-driving harness, should be left untightened on a horse. The reason is that the landscape outside of the arena, where this harness was designed to be used, is often hilly. Hilly landscapes don't always offer pathways to travel that are straight up or straight down the slope. Often the line of travel goes across the slope to some degree. When you look at a picture of a driving horse coming at you across a sloping field, you see clearly that the vehicle is tilted for the slope while the horse's body remains plumb. The articulation of the joints in the lower limbs of the horse allows him to stay vertical, even on the steepest slopes.

When the axles of the vehicle are tilted because of the slope, this also tilts the shafts on either side of the horse. If your girth is fastened as tightly as you fasten it on your riding horse, there is going to be a powerful wrenching effect on your horse's body. If the girth is loose enough to circle around the horse's body to traverse hills, the tug loops continue to do their job of holding up the shafts. At the same time, the whole surcingle shifts around on the horse's body without painfully wrenching it.

On most well-made pleasure-driving single harness, the uptugs that hold up the breeching are designed to pass easily through a slot in the back strap, which is done that way for all of the reasons above. The shaft wraps follow the shaft upward on the uphill side of the horse and downward on the other side.

Chapter Two

Collars and Their Use

When I head out to start the haying operation on our farm, you can be sure that one of the things I am thinking about most is the collar fit on my team. I know how easy it is to misread how a collar is fitting and then inadvertently give my horses sore shoulders or necks. Even when the collars fit perfectly, you can still end up with collar sore problems if your horse is hitched to a vehicle or implement that was not designed properly. Two of my mares got sores on the top of their necks one afternoon from a hay rake with a heavy tongue. The New England D-ring harness is the only type of work harness that addresses this problem by transferring tongue weight to the backs of the horses. Another time, I started a collar sore on a fat horse's neck while he was just standing tied to the fence for an hour! The collar fit a bit too snugly along the sides and I had failed to pull the mane hair out from under the collar. When I pulled the collar off, the skin under where the mane hair and collar had been was irritated. I have learned a lot from episodes like that. Now I watch closely and pay attention to how collars fit and how the horses are worked to avoid such problems.

WHERE DOES A COLLAR FIT?

In general, collars need to fit slightly ahead of the horse's shoulder for a good, safe fit. When the collar is pushed back on the horse's shoulder, there should be just enough room for an average-size person to fit the flat of their hand inside the throat of the collar. Be sure to note that the flat of the hand should fit horizontally, not vertically. I think that many horsemen fit their collars too loose. Surprisingly enough, more horses get sores from a collar that is fit too loose than from one that is fit too snug.

I try to fit the collar so that when in draft, it sits against the muscle on the leading edge of the shoulder bone (the scapula), called the supraspinatus muscle. The scapula has a bony ridge down its middle that divides the supraspinatus from the infraspinatus muscle. The bony ridge of the scapula is not covered by much muscle, only by skin and nerves, making it a very painful place for a collar to press against. This ridge has a nerve, called the suprascapular nerve, that runs over the top of it, waiting (or so it seems) to be crushed by a collar that is too big and gets pushed onto the nerve with a heavy load.

▼ **The bone structure and suprascapular nerve.**
It's some unfortunate placing to have the suprascapular nerve directly over the top of the ridge of the scapula, where it can be easily damaged. Collars that are too big or too loose end up crushing the nerve if the horse pulls hard on something.

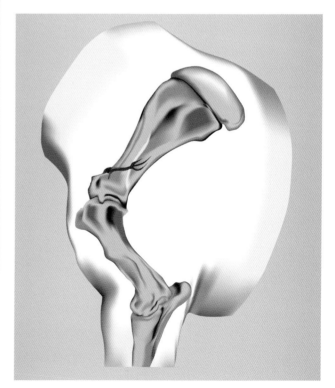

▲ **You want a proper fit.** It's hard to reach perfection, but this is a fit that looks pretty close. On a draft horse, you should be able to fit the flat of your hand into the space between the neck and the collar at the bottom of the collar. The sides of the collar should fit neatly against the hair, neither pinching nor sloppy.

SWEENEY

Crushing the suprascapular nerve leads to a condition called sweeney, which is when the supraspinatus and the infraspinatus muscles atrophy (shrink). When you see a horse with sweeney on one side, you know it was probably caused by an impact injury. It is often caused when a horse gets kicked there, or when a group of horses try to squeeze through a gateway together and the one on the edge gets slammed into a post. When you see sweeney on both sides of a horse's neck, you might suspect poor collar fit. Draft-horse people get duped into using too large of a collar because many of them are caught up in the vanity of having a big horse. They brag about their horse's shoe size, height, weight, eating habits, and, unfortunately, their horse's collar size, so there is a tendency for some horsemen to fit their collars too big for a very poor reason.

COLLARS AND THE CENTER OF GRAVITY

Most collars have an extra layer of leather padding the spot on the forewale (the front, outer surface of the collar; the back side of the collar that goes against the horse is called the afterwale) where the trace attaches. This special spot is called the point of draft of the collar. If the point of draft is too low because the collar is too long, the trace will be carried too low on the horse's body and make the load seem to pull downward on the horse. In other words, the horse has a hard time getting behind the load it is pushing because the load is being carried below the horse's center of gravity.

Around the show ring, I notice that many of the horses shown are in collars that are much too big. Besides the bragging issue, another factor operating there is that those fancy collars and the housings and hames that fit onto them are very expensive to buy. Younger, undersized horses are probably fitted with collars that are too big so the owner can avoid buying collars that the horses are going to grow out of. A further reason for overly long collars on show horses might be that a horse's only way of compensating for a collar that puts the point of draft too low is for the horse to lift his head and neck. A high head is highly regarded in the show ring because of the belief that it elevates the front end of the horse and causes more animated hoof action.

MEASURING THE LENGTH OF A COLLAR

When you order a new collar, you'll be presented with the choices of length and shape. Length is simply the measured length of the opening in the center of the collar. That length is a whole number of inches when the collar is made (19, 20, 21 inches, etc.). If you want a collar that fits your horse well, it is best to try on a few collars of different sizes and measure the result before ordering a new collar for your horse. Even if the collars available aren't a perfectly sized match, you will get a much better idea of what to order this way compared to holding a yardstick up to a horse's neck and eyeballing it.

▲ **The regular-cut collar is also sometimes called a full-face collar.** This collar is made relatively narrow and is the right shape for horses that are in hard-working condition and for younger, slimmer animals. This was the most popular shape for collars back when using horses for farming was common. Notice how the white buck stitching follows the edge of the collar.

THE RELATIVE SHAPE OF COLLARS

Horses' necks come in various shapes, so collars also come in shapes to match those necks. Regular-cut, or full-face, collars are made for narrow-necked animals. Many mules, mares, and geldings that are being worked hard and younger animals wear regular-cut collars.

Half-sweeney collars are made to fit horses with a bit more roundness to the top of their necks. Fatter, underworked mares and geldings, and older animals, often use half-sweeney collars. Collar manufacturers say that before the invention of the tractor, most of

▲ **The half-sweeney collar is the most popular shape of collar for modern-day horses.** It is designed for horses that are carrying a bit more weight and thus have a wider, rounder shape to their necks. Notice how the white buck stitching scallops away from the trailing edge of the collar near the top. There is less stuffing in that area. Full-sweeney collars are so rare that we couldn't find one! They are designed for stud-necked horses that have very round-shaped necks. Full-sweeney collars have even less stuffing at the top of the collar and have a more pronounced scalloping of the stitching.

the collars in use were regular-cut collars. Now most of the collars made and sold are half-sweeney collars.

Full-sweeney collars are reserved for the seriously obese, the "stud" neck, and for the horses that have been fully sweeneyed. Sweeney causes more atrophy at the lower portion of the neck than at the top, leaving the shape of the neck more relatively round. Of the three shapes, full sweeney is the shape you see the least.

Collar manufacturers attain the various shapes by varying how much stuffing they put into the top portions of the collar. Regular-cut collars have no stuffing relief at the top of the collar. Half-sweeney collars have the stitching scalloped near the top of the collar. Full-sweeney collars are stitched so that the top of the collar has very little or no straw stuffing at all; the stitching scallops in so that it is very near the hame groove at the top of the collar.

COLLAR CAPS

The top cap of a collar is usually made of a very heavy, shoe-sole type of leather that has been formed to the approximate shape of the top of a horse's neck. Usually the top of a collar is bound together with a buckle and billet of leather. Some bind the opening together with an over-center buckle that is very similar to a rubber overshoe buckle. When manufactured, collar makers hold the collar together with a short piece of leather strap that is riveted to join the top of the collar together. Manufacturers expect the purchaser to use a knife and cut the short piece of strap so the collar can be opened.

Some collars are made to be fully adjustable—in other words, they fit horses of about a 3-inch range of collar length because of the expanding cap at the top of the collar. If using a collar pad, these seem to fit horses very well. They can be a little flat on top when used without a pad underneath and sometimes are more prone to causing soreness at the top of the neck, especially if the harness and the vehicle put too much weight there.

Some pleasure-driving and show-ring collars don't open at the top at all. These are called Kay collars, after the person who patented them. To get them on, it is customary to stretch them laterally between your hands and one knee until they are wide enough to fit over the head of the horse.

▲ An adjustable collar top is also sometimes called a "sliding-top collar" or "expanding-top collar."
This one is made to fit a 3-inch range of collar sizes, depending on how the top is adjusted. Some teamsters think these adjustable collars are not so good because the top is too wide and flat to match the top of the neck on some horses. It seems to work well if you are using a thick collar pad on the collar. Collar tops open and close with a variety of fastening arrangements. The other collar tops shown here are buckle and billet. Some collars are not designed to open at all. Being able to open a collar helps a lot, especially on horses with heads that are bigger than their necks.

▶ Full collar pad. Modern full collar pads are offered in new materials that feature a wide variety of color availability, foam padding that is protected from getting soaked by sweat, and a heavy-duty truck tarp material against the horse's shoulders to cause sweat to run onto the horse's front legs. Older ticking-surfaced pads were made to absorb the sweat into the padding material, which is usually deer-hair felt. The problem is that sweat was made to cool a working animal as it evaporates. Sweat that gets absorbed into a collar pad just makes the pad heavier because the padding loses its springiness from being soaked.

▲ Top pad. If your collar fits well along the sides but it is just a little too long, you can use a top pad to shorten the collar length by about an inch.

▲ **Collar pad on collar.** This collar pad attaches to the rim of the collar with spring clips. It is not possible to open the collar with the pad attached. Usually the left side of the pad is unclipped from the collar when it is opened to put it on or take it off the horse.

COLLAR PADS

I used to always say that collar pads are meant to be shims that fill in loose or otherwise poorly fitting collars. Collar pads are made of different kinds of material that is of varying thickness and can be made to pad the entire collar (except at the throat) or only certain portions of the collar. Sometimes top pads are needed to lift up a collar that is too long. Other times, side pads are all that is needed to fill in where a collar fits too loosely along the sides. However, after experiencing creating sores on my horses, I now size collars to be fit with a thick collar pad for heavy work horses. I think it's probably a lot like wearing socks in your shoes. I've noticed that virtually all of the horse-pulling competitors have their horses in full collar pads. When ordering collar pads, measure your collar and add 2 inches to the collar length for the proper pad length.

TOUGHENING UP THE SHOULDERS

My father told me that many of the old-time farmers would wash their horses' shoulders down with salt water to toughen up the skin around the shoulders and make it more resistant to soreness. I have never done that to my horses. Instead, I try to gradually increase the workload over a couple of weeks' time, especially after they've had a long time off. I've never seen a callous on a horse's neck; I'm not sure a horse's skin is capable of such things. On a fit horse, the skin at the shoulders is the same as an unfit horse's skin, or at least it seems that way to me.

It also seems that the key to avoiding sores from working a horse day after day is to pay close attention to collar fit. As the growing season or the wagon train moves on, day after day, the constant use will cause a certain amount of change in the body shape of the animals being worked. It'll be a gradual change, but it is the sort of thing that you will need to pay attention to in order to avoid hurting your animals.

MAKING IT FIT

If your horse's collar doesn't seem to fit just right, it's okay to soak it a little and then put it onto the horse for some light work. This is somewhat like the practice of soaking a new pair of leather boots in water, then wearing them until they dry in order to get a good fit.

Be careful to not soak a poorly fitting collar for too long or else you could soak into the straw that is under the leather. If you simply throw the collar into the horse tank and put a cement block on top to hold it down, you'll end up with a dripping wet collar that weighs about 50 pounds.

It is better to use wet towels to soak the surface that goes against the horse and test the softness and weight of the collar during the soaking process to prevent water-logging your collar.

OVER THE HEAD?

I train all of my horses to have the collar put on and off over their heads. This method causes a lot less trouble, especially if there is a collar pad involved. Notice that some horses and mules have heads larger than their necks, so the collar won't go over their heads. When I buy a team, I always ask if the horses are used to having the collars go on over their heads. Horses that have never had the collar go over their heads will get plenty concerned about it at first. Like anything else, if you go slowly and make sure the horse isn't being hurt as the collar goes over his head, he'll soon duck his head and help push it over his eyes when you put the collar up to his nose. "Over the head" training will go smoothly if the horse is not tied up the first few times you do it.

KEEPING COLLARS

Try to accumulate various collars in the size range of the horses that you are using as a means of having a better chance of having the right collar on hand when harnessing a new horse. Collars keep well if they are hung up properly (by the throat, not the cap) and if they are never broken by allowing them to open too far. Collars are stuffed with straw to get and maintain their shape. The extra-tall variety of rye straw works well to maintain that shape for the intended use, but it is not very resistant to breaking. When an uninformed helper opens a collar and allows one side to flop toward the ground as the other side is lifted up, the straws at the throat of the collar are broken. Such misuse will soon make a new collar feel like a limp dishrag. Collars and collar pads bridge the gap between the soft flesh of the shoulder and the hard steel or wood of the hames in a very particular way. Using a collar with a flaw will not work.

MEASURING THE DRAFT

Generally, collar makers charge more for collars that have more stuffing put into them. The thickest diameter of a collar will be at the "point of draft" area, the place where the trace attaches to the hame. Collar makers advertise the draft of a collar, which is a measure of the girth of the collar at that point. Usually, the heavier the load to be pulled, the more comfortable the load will be if the collar has more thickness at that point. Pleasure-driving and lighter work collars get by with less thickness here because the horses are not pulling as much weight.

HAME MEASUREMENT AND FITTING

Draft-type hames have a hame-adjusting ratchet at the top that allows the hames to be fit to collars of various sizes. On modern hames, the ratchet adjustment has three notches that are spaced 1½ inches apart, allowing for 3 inches of adjustment between the top and bottom notches. Harness makers refer to hames by their longest available notch, so a 24-inch hame is measured from the top notch to the bottom of the hame, not including the loose link on the bottom. The bottom notch on that 24-inch hame makes the effective length of the hame 21 inches, and it would fit onto a 21-inch collar when used in that position.

▲ **Measuring "Draft" of Collar.** Collar makers often refer to the draft measurement of the various collars they sell. Generally, the more stuffing a collar has at the point of draft, the better it is for heavy pulling. This collar has a draft measurement of 19 inches.

▶ **How to measure hames.** Draft hames are made to adjust over three different sizes to match four different collar sizes. Harness makers refer to the longest possible measurement when referring to hame size. If these hames fit 21-, 22-, 23-, and 24-inch collars, they are referred to as 24-inch hames. The notches on the hame-adjusting ratchet are 1½ inches apart, with the top notch at 24 inches as shown here. The bottom notch is at 21 inches.

It is important that hames fit into the hame groove of the collar so there are no visible gaps along the length of the hame. If you start by measuring the length of the collar, you can then decide which link on the hame corresponds to that collar length. Collar and hame length should match exactly, but if that is not possible, it is better to use a hame setting that is a little shorter than the collar. The top hame strap is used for width between the hame's adjustments, and it also can be let out enough to get extra hame length when needed. If the top hame strap is too short, pulling the hames into the groove of the collar with the lower hame strap will cause a painful pinching action at the top of the hames.

Some teamsters make quite an effort to pull their bottom hame strap super tight, to prevent the hames from pulling out of the groove of the collar. If you have a problem with your hames pulling out of the collar groove while working, the usual cause is that the harness is not properly adjusted or the horses are not being hitched properly. The breast strap comes too far forward when the horses stop and it puts forward pressure on the hames when stopping. This movement pulls the hames out of the groove.

▲ Measuring the bottom notch on a 24-inch hames.

▲ A 21-inch hames on a 21-inch collar.

▲ **Improperly fit hames.** If you can see large gaps between the hame and the collar, you need to keep working on the fit adjustments. Improperly fit hames cause a disruption of balance between the collar and the hame, which might affect the comfort of the horse. In this case, the hame ratchet was at the right setting, but the top hame strap was too tight.

▲ **Hames fit into groove.** Between the rim and the back of the collar is a deep groove that the hames are designed to fit into. The hames should be adjusted with the top and bottom hame straps and the ratchet setting so the hames fit neatly into the groove.

Chapter Three

Making the Connection

If you entered the hobby of driving horses without growing up in the culture, you'll need to learn about some of the connecting considerations between the harnessed horse and vehicles. There is a lot of variety in this subject, but we're keeping it simple and starting with connecting devices for the single horse, then progressing up through some of the more commonly seen multiple hitches. The measurements provided will allow you to compare what is in these pictures with other vehicles and implements you will encounter. As with the harness section in Chapter 1, your emphasis should not be so much on memorizing exactly what each of these pieces of equipment look like and what they measure; you should focus more on how they function and why they work that way. With the "how it works and why" approach, you'll be ready to make good connections between the harnessed horses and your equipment.

▶ **Singletree.** An option for hitching a single horse in draft-type harness to an implement that drags along on the ground (providing it has its own brakes without need of shafts) is shown here. The singletree provides a way to attach the heel chains of a draft-type harness to a load, and the attaching ring provides a central pivot point for the pulling comfort of the horse. Shoulders take turns being in front as the horse strides. The singletree allows for free shoulder motion. This is essentially the connecting set-up for all things that drag along behind a single horse.

▲ **Draft harness attached to singletree.** Usually when a horse is hitched to anything that drags along on the ground like this, the singletree is hooked into the longest possible link on the traces. When the traces are brought low to drag a load that is low to the ground, the extra length at the end link is needed to provide clearance for the horse's striding rear legs. Sometimes traces are shortened by attaching to shorter links when more of a lifting motion is needed.

◀ **Common horse cart.** This is a side view of a fairly typical horse cart designed for pleasure driving. This particular cart is a reproduction of an older cart that was called a Kentucky breaking cart. The tall wheels, tilted seat, extra-long shafts, lack of backrest, and standing platform behind the seat are all features that make this an ideal training vehicle. The tall wheels get the driver's eyes elevated, so he or she can see the horse's ears instead of just the hind end. The tilted seat and no seatback simulate the driver's wedge that most high-performance competition vehicles provide. The driver is on his feet nearly as much as on his hind end, very similar to the feeling when riding a horse. The lack of seatback greatly frees up elbow motion, which is often needed for driving young horses. Extra-long shafts put the horse out of kicking range of the front of the vehicle, and the wide step at the rear is a safer place to drive from for the first drive or two. It also makes a nice place to carry along extra passengers when traveling.

▲ **Singletree on a cart.** This singletree on a cart is designed to work with pleasure-driving harness traces. Pleasure-driving traces have slots punched into the ends of the traces for length adjustment when hitching. One of the slots punched in the end of the trace is pushed over each end of the singletree, where it is secured in place with a spring-steel clip that holds the trace. Singletrees on vehicles allow for free shoulder motion on the horse because the singletree pivots in the middle. Without a singletree, the horse's shoulder area receives a good scrubbing as the hide scuffs against a collar or breast collar as the horse moves.

▲ **Trace extenders.** Trace extenders are an inexpensive way to extend the length of traces as needed for certain vehicles. Most pleasure-driving vehicles made for saddle horse–size horses have shafts that are about 7 feet long. Most pleasure-driving traces are also about 7 feet long. If the shafts on your vehicle are 8 feet long, as they are on this training cart, trace extenders are needed to put the horse where he needs to be in relation to the shafts. Another nice thing about trace extenders is that they easily allow for connecting a horse in draft horse harness (harness with heel chains) to this pleasure-driving vehicle. The elongated links of the heel chain are pushed through the buckle on the trace extender and fastened with the tongue of the buckle. Trace extenders also easily attach to pleasure-driving traces.

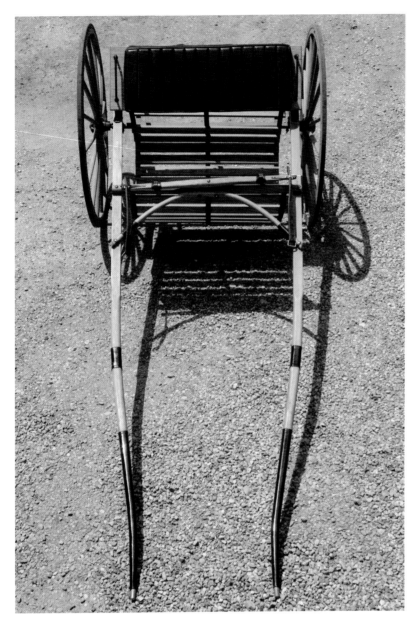

◀ Shafts. These shafts are made of steam-bent hickory wood and are limber enough to handle all sorts of equine antics, such as when a horse rears up and turns at the same time, coming down with one shaft over his back and the other shaft under his belly. They snap back into place as soon as the horse starts going forward again. Shafts on this vehicle hold up the front of the cart and allow for turning that follows the movement of the horse. It also provides a horse-activated braking system that automatically keeps the vehicle from getting closer to the horse. The extra-long shafts on this cart are 8 feet long from singletree to shaft tip. This cart has shaft tips that are 26 inches apart at their narrowest point, a foot or so back of the tips, and this spacing works well for 14- to 18-hand horses. Many draft-horse carts have a shaft tip spacing that is about 36 inches at its narrowest point. This looks awful and seems a little dangerous with the shaft tips being a true menace to animate and inanimate objects near the path of travel. At the rear end of the shafts, the spacing is 42 inches apart. The extra width at the rear of the shafts allows for more freedom of movement for the horse, especially through turns. In a sharp turn, a horse's hip will usually be pushing into the shaft that is on the outside of the turn. It is interesting that many horses in a tight turn in shafts seem to point their tails in the direction of the turn like a turn signal.

▶ Footman's loop. Shafts need to have some sort of loop on the bottom of each shaft to activate braking. Holdback straps, fastened to the breeching of the harness, are run through the footman's loops and wrapped around the shafts. The straps are wrapped in a special way that relies more on the friction grip of the wraps than the little screws that hold a brass footman's loop onto a wooden shaft. On this cart, the leading edges of the footman's loops are screwed on 36 inches from the shaft tips. This footman's loop placement seems to work well for a range of horses, from large pony to full-size draft.

MAKING THE CONNECTION FOR TEAMS

One or more pairs of horses wearing draft-type harness are called teams. A characteristic of the hitching of a team is that the horses are separated by a pole or tongue. The tongue running between the horses is there to steer the vehicle and activate the braking system of the harness, automatically keeping the vehicle out of the horse's space. The neck yoke is the connecting apparatus that holds up the end of the tongue, steers the end of the tongue, and provides the braking connection between the harness and the vehicle.

The pulling of the vehicle is accomplished by attaching the traces to singletrees that are a part of a set of doubletrees. The doubletrees assure even distribution of pulling duties. They pivot in the middle, which allows one horse to be slightly ahead of the other (if needed) while still keeping the pull on each trace relatively even.

◀ **Team hitched to a harrow.** Many farm tools don't need a tongue because they drag so much. This double section of spring-tooth harrow is an example of an implement that doesn't usually need a tongue. When hooking to this type of implement, the traces are all that needs to be attached because the implement can be steered and pulled forward by traces alone. When hitching to this sort of thing, the traces are usually fastened into the longest available link on the heel chains. Notice how the lazy straps on the harness keep the traces lifted up when the horses are stopped. They help keep horses from getting their back legs tangled in the traces while swatting flies or fidgeting. The two-horse evener bar behind the singletrees assures equitable effort, even if one horse lags behind.

◀ **Doubletree on sled tongue.** On vehicles with tongues, the doubletrees are usually attached like this. The clip extending over the top of the draft hole on the doubletree helps relieve shearing of the doubletree attaching pin by securing it at the top. The spacing between the two singletree mounting holes (on center) is referred to as the effective width of the doubletrees. In this case, they are 42 inches apart.

▲ **Side view of sled tongue.** This training sled tongue has an overall length and an effective length. When hitching horses, we are most concerned with the effective length, which is the distance from the draft hole (where the doubletrees attach) to the forward side of the neck yoke stop at the front end of the tongue. When I first started hitching draft horses 30 years ago, most tongues were 8 feet 6 inches effective length. Now most tongues are about 9 feet 6 inches or longer in effective length because horses have gotten that much longer. If you buy an antique vehicle or implement, be sure to see whether the tongue length is long enough for your horses.

◀ **Neck yoke on sled tongue.** This slip-on (and off) neck yoke is positioned on the end of the sled tongue as it would be for hitching. The neck yoke holds up the end of the tongue, steers the sled right and left as the horses move in those directions, and, if needed, it provides a braking attachment from the coupler snaps on the harness to the vehicle. When used on dirt or grass, the best use of a neck yoke and tongue is to keep the horses from backing into the sled, which sometimes happens with green animals on sleds without tongues. Notice the added-on chain to keep the neck yoke from slipping off. The effective length of the neck yoke is the distance measured between each end ring attaching point. In this case, it is 42 inches.

TONGUE (OR POLE) ENDS

A major feature of hitching apparatus is the tongue, or pole, of the vehicle. Most tongues on vehicles have a mounting place for the doubletrees near the vehicle and a mounting place for the neck yoke at the far end of the tongue. There are a variety of different ends that are used on tongues, depending on whether the person who designed the tongue was striving for convenience or safety.

Standard Stop on Tongue End

Most pole ends look like this. They were designed for a neck yoke with a big ring in its middle to slip easily onto the end of the tongue. It should rest against the stop or protrusion on the bottom of the tongue when in use. The big problem with a tongue end like this is that the neck yoke, which does so much for steering and brak-ing of the vehicle, also easily slips off this type of tongue

design. If a trace comes loose from the doubletree or if the teamster made a poor decision when hitching (more about this later), the neck yoke can slip off the end of the tongue and send it crashing to the ground. At that point, the horses have all of their ability to pull the vehicle, but they have no ability to stop or steer it. If your vehicle has a tongue end like this, you can make sure the neck yoke stays with the tongue by chaining it to the tongue with a screw link connection.

Bolt-On Neck Yoke

A bolt-on neck yoke is more often used on implements than vehicles, but it is a safe and secure way to make sure that the neck yoke and the tongue end don't part com-pany. The nut on the bottom of the

neck yoke bolt should be double-nutted, pinned, and peened or it should be an aircraft type that won't rattle off. The steel end on this wooden tongue is a nice way to effectively lengthen a too-short tongue.

Safety Stop

This tongue allows you to slip on the neck yoke fairly easy, but it won't let a neck yoke slip off while it is in use because of the steel piece welded on at the end of the tongue.

Tractor Tongue End

This type of tongue end allows many different uses for the vehicle. It can be hitched to a tractor or to a pickup if it is needed for long-distance hauls. A bolt-on neck yoke can be bolted into the hitch pin hole or a slip-on neck yoke can be pinned securely into place.

MAKING THE CONNECTION FOR MULTIPLE ROAD HITCHES

Many people's first exposure to driven horses is at a major event or at a big draft-horse show. At such places, horses are often being shown in multiple hitches in an arena. These are technically road hitches because they are multiple hitches that are arranged for travel. The other type of hitching that uses more than two horses is the field hitch. The main difference between the two kinds of multiple hitches is that the set-up for road hitches strives for narrowness to allow easy passage in traffic situations. This keeps the horses arranged so they are no wider than the wagon being pulled, but they also can be strung out ahead of each other for a fair amount of distance when you are using a large number of horses. In field uses of multiple hitches, there's no need to be worried about width.

Hitch wagons for hitching multiple horses come equipped with a fifth horizontal wheel at the pivot point of the front axle. This wheel is a great safety device because it allows a team to turn very sharply while all four wheels maintain contact with the ground. Four-wheeled vehicles cannot turn as sharply, which presents a danger if the team is out of control and takes a sharp turn. When the tongue reaches its limit of turn, it will begin to break or bend and the vehicle can overturn. Most fifth-wheel vehicles have a stiff tongue that keeps the forward end of the tongue held up without any effort by the team of horses.

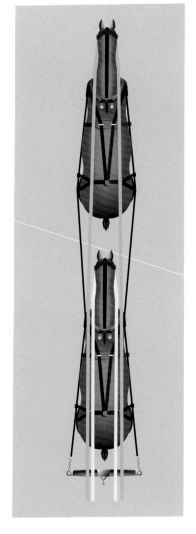

◀ **Tandem hitch.**
A tandem hitch is one in which two horses are driven with one ahead of the other. This is the way that British teamsters seem to prefer using two horses, instead of putting the horses side by side. When driving from the vehicle with lines going to both horses, the skill level required is very similar to that needed for driving a four-up of horses. Those who have tried this hitch say that this can get pretty exciting because single horses in the lead position are not quite as steady as having a pair up there. Three horses strung out like this is called a random hitch.

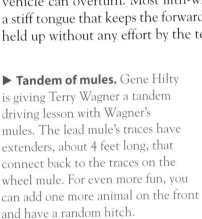

▶ **Tandem of mules.** Gene Hilty is giving Terry Wagner a tandem driving lesson with Wagner's mules. The lead mule's traces have extenders, about 4 feet long, that connect back to the traces on the wheel mule. For even more fun, you can add one more animal on the front and have a random hitch.

◀ **Hitch wagon with stiff tongue and fifth wheel.** When the end of the tongue is held up by a stiff tongue, pole chains are a simple way to transfer stopping and steering from the harnessed horses to the vehicle. The fifth wheel lies horizontally at the pivot point of the front axle and allows almost complete freedom for turning this vehicle. In an out-of-control, too-sharp turn, the horse on the inside of the turn would be able to turn all the way around to touch the rear wheel on the inside of the turn. I've seen horses hit the rear wheel with enough force to bump the rear of the wagon without actually unbalancing it or stressing the tongue.

▲ **Spreader bar.** At the wheel position, the pole chains are stabilized by a bar with elongated loops welded to the end, which is called a spreader bar. The elongated end loops pass through links on the pole chains before they are snapped into the coupler snaps on each wheel horse's harness. The spreader bar keeps the horses' distance apart set to match the doubletree width, and it also assures that the stopping and backing of the wagon keeps the horses' bodies straight instead of pulling them toward the tongue end.

▲ **Spreader bar ready for hitching.** In this photo, you can clearly see the hitching apparatus at the end of a typical hitch wagon; the spreader bar is ready for hitching.

▶ **Unicorn hitch.** A unicorn hitch is a nice way to add quite a bit of horsepower without getting too wide to go down the road. The unicorn hitch drives a lot like driving four up.

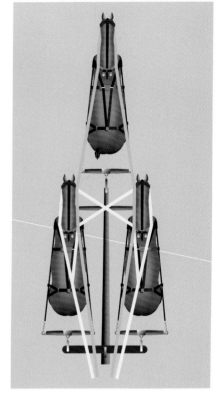

▲ **Singletree added on for unicorn.** On a hitch wagon, the singletree needed for the unicorn hitch is attached to the top loop on the tongue end by a chain. The leather keeper strap closes the end of the top loop to make sure the singletree stays in place.

▶ **Four up.** For road hitches, four horses driven in two teams of two is called a four-up. The horses near the wagon are called wheelers and the horses out in front are called leaders.

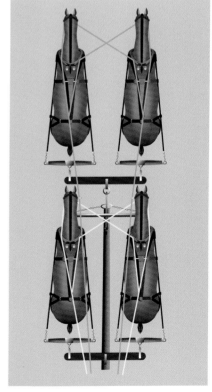

▲ Four-up equipment. To add a team on the front of the wagon, all that is needed is to add on an extra set of doubletrees. These are sometimes called lead bars. They are made lighter than standard doubletrees.

▲ Hitch wagon set up for hitching a six-up.

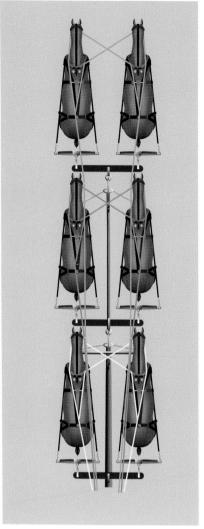

▲ Six-up. In this configuration, there are wheelers and leaders, and the middle team is called the swing team. Each team is driven with coupling lines that have enough length to reach back to the driver. On draft hitches, the lines are passed through special rings on the inside (medial side) of the bridles on their way back to the driver.

▲ **Detail of the connections between wheeler's tongue and swing pole.**

▲ **Detail at the far end of the swing pole.** Another set of lead bars are attached to the end of the swing pole for leader attachment.

▲ **Rein guide on a swing team bridle.**

▲ **Gene Hilty is driving a six-up of Belgians at a draft-horse show in California.**

MULTIPLE FIELD HITCHES

Back in the 1920s and 1930s, the United States Department of Agriculture tried to figure out a way for horse farmers to compete on a scale with the tractors that were becoming so popular. The USDA developed a lot of information and traveled around the country holding field days at which multiple field hitches for draft horses were promoted and explained.

Much of the line (rein) handling for the big hitches featured simplicity for the farmer. This had to be an activity that compared favorably with the simplicity of turning a steering wheel on a tractor. There was much use of jockey sticks, buck-back reins, and lead straps (or ropes or chains) to easily control multiple horses.

One of the selling points for using horses instead of tractors was that horses in big hitches could be used together for heavy activities like plowing and discing, and the next day they could be split into multiple teams for wagon work, logging, and lighter farming activities. A farmer who invested in a tractor big enough to plow with three bottoms had to use the same steel-wheeled, big, heavy thing to mow his hay and haul a wagonload of corn to town. He didn't have the versatility of a horse farmer.

In field hitches, there is great effort taken (lots of mathematical figuring) to assure that every horse in the hitch is theoretically pulling the exact same amount as his buddies, even if he is stepping ahead of or falling behind his teammates. With road hitches, there is usually not so much concern about equity, because effective driving of multiple hitches involves having some horses pull more while others pull less at certain times. This is accomplished by having a teamster who is able to set some of the horses back "out of the traces" while at the same time having others do more of the pulling, thanks to a skilled use of the driving lines. For example, when turning sharply, you don't want your leaders to be pulling very hard on the traces because if they do, they will trip the other horses that are further back in the hitch.

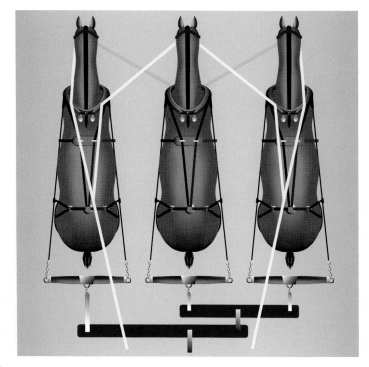

▲ **Moving three abreast with buck straps.** This arrangement for three abreast uses an ordinary set of team lines that are adjusted for the narrow spacing of a three-abreast evener (the horses are usually about 30 inches apart). The center horse is the only one truly being driven by the coupling lines that go back to the driver's hands. The buck straps anchor to the upper or lower hame rings on the center horse and attach to the medial bit rings of the side horses. The type of three-horse eveners shown here assures even pulling by all three horses, even if one gets a little ahead of the other two.

▲ **A team moves three abreast.**

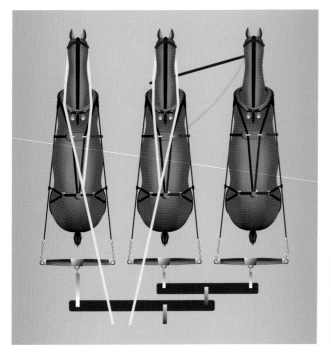

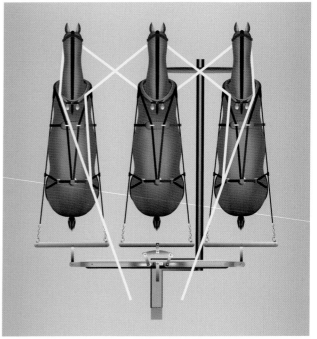

▲ **Moving three abreast with a jockey stick.** This is a simple way to handle the steering and stopping of three abreast for field use. The stick horse has a jockey stick that anchors on the hames of one of the other horses and goes to the outside bit ring. The side of the bit that is near the other horse has a strap or lead rope that anchors back to the bellyband billet/trace junction on the same driven horse. The two with the regular team lines are the only ones that truly feel the teamster's hands.

▶ **Move three abreast on harrow.** You can easily see how the three-horse evener scissors when one horse lags behind, making his pull stay even, regardless of where he is putting himself. You can see a wave of energy heading up the left line toward the left horse's hind end. Most teamsters communicate a need for more energy in individual horses this way. It is also possible to carry a whip to handle the problem, but it would have to be a long one. You don't often see a farmer carrying a whip because sending energy up the line like this works pretty well without having to carry extra equipment. The safest place for a teamster is behind the harrow, as shown here. Most horse-drawn farm equipment that is well designed seeks to place the driver behind equipment for safety.

▲ **Moving three abreast with offset tongue.** On vehicles and implements that roll freely forward, you sometimes see an offset tongue used with three abreast. This arrangement pulls the load with three horses and holds the load back; it also steers with just two of the horses, which is usually adequate. I've used this hitch on our manure spreader when the going gets too muddy for a team.

Chapter Four

Hitching Considerations

In this chapter, we take a look at the considerations a person makes when attaching the harnessed horse to a vehicle. Single hitching (putting to) and putting away is covered first, followed by hitching and unhitching of a team in draft-type harness to a vehicle that has a tongue. There is a lot of detail in this chapter because the information determines how safe and comfortable the driving experience will be for your horses.

Once you have digested the instruction contained in this chapter, your once-innocent appreciation of the sight of horses being driven past you at an event will never be the same for you. After this, you'll know at a glance if the person on the reins has the good sense to know and care about hitching considerations, or if they are just struggling along doing things because of some half-baked idea about what's right. Once you see how things are supposed to be adjusted when hitching, your new understanding will make it possible for you to be more considerate about the comfort of your horses and the safety of all in the future.

▲ **Hold shafts high.** With vehicles that have shafts, the usual way to hitch is to bring the vehicle to the horse, as shown here. By lifting the shafts high, the narrowest part of the shaft shape is brought past the wider hips of the horse, making it easy to get the shafts to the right position relative to the horse. Most people who are hitching single tend to use a helper so that there is a person on both sides of the horse carrying out hitching activities. The driver of the horse is the one calling the shots and assuring that the helper is doing things well on the unseen other side of the horse. I like to have my horses trained to be hitched while tied, but some whips seem to have a great fear of this. It's okay to employ an extra helper to stand at the horse's head to hold the horse while being hitched. Just be sure that the person holding the horse's head knows that once the horse is in shafts, it is no longer okay to stop a horse that is moving ahead by having the horse circle around the person with the lead rope. When the shafts are along the horse's side, a forward-moving horse needs to be stopped with straight rearward pressure so the horse isn't thrown over onto the outside shaft.

The steps outlined here are what I would call conventional. This seems to be the hitching sequence used by the majority of horse people who do these things well. It is not just my own private interpretation of what I think is right. If you learn the hitching and unhitching steps shown here, you won't be left out of the loop if you are helping a knowledgeable horse person from another part of the United States or even another part of the world.

The steps taken when hitching and unhitching is a subject that a person memorizes. It is through memorization of these steps that a person fluidly moves from one attaching point to the next without making a mistake. When you are in the company of other people while hitching, you'll make a good impression and show off your skills if you are always ready with the next step at the right time.

▲ **Put shafts through tug loops.** After the shaft tips are brought forward near the surcingle, they can be lowered and put through the tug loops. If you are hitching a horse in shafts by yourself, you can stay on the left side as you bring the shafts to the horse, then reach over the horse's back and lift the uptug on the tug loop so the off (right) shaft can be put through the tug loop. Once the tug loop is over the off shaft, both shafts are lowered. The shafts are then brought forward a little more in preparation for hooking up the traces.

▶ **How to position shaft tips.** There is a wide range of approved options for where to position the shaft tips relative to the horse. In combined driving competitions, the shaft tips are placed just forward of the tug loops when doing the marathon phase, which leaves the shoulders free. The forward approved limit for shaft tip position is at the point of the shoulder. In draft horse shows you often see drivers hitched with a foot or more of shaft tip sticking out in front of the horse's shoulder, but let's not do that. In most cases, it works well to have the shaft tips alongside the horse's shoulder. This placement allows a solid and safe place for a horse to touch a shaft when turning. I like the forward limit of shaft position to be a couple of inches behind the point of the shoulder. Being back some from the shoulder prevents the shaft tips from shafting fence posts or innocent bystanders with the outside shaft as the horse rounds a corner. A likely looking slot in the trace is chosen (the same slot is used on both traces, so the trace length stays even) and tested by putting the

vehicle into draft. This means the vehicle is rolled to the rear with enough pressure to take the slack out of the traces, and it stops the shafts at their most rearward limit of travel. When choosing the ideal place to connect the horse to the shaft tips, I like to have the horse slightly uphill from the vehicle so it is easy to keep the traces tight and the vehicle in draft when viewing shaft tip positioning.

▲ **Attach traces.** Because these shafts are extra long, trace extenders are being used. The buckle end of the trace extender easily buckles into the slots on the trace end.

▲ **Shaft tips when in draft.** This is a very acceptable shaft position for general-purpose driving.

1

▲ **Put holdbacks through footman's loop.** Starting with the holdback strap hanging limply by the horse's side, medial to (deeper than) the trace and the shaft, pick up the end of the holdback and pass it through the footman's loop. Be sure the rough side of the leather strap is facing up as it comes through the loop and that the holdback strap is below, not above, the trace.

2

▲ **Wrap forward.** Remove any excess slack in the strap and begin to wrap forward of the footman's loop and only around the shaft. The finished side of the holdback strap is exposed to see where it wraps around the shaft. Don't wrap around the shaft and the trace. You want to avoid tying the trace to the shaft because it can cause sore shoulders on your horses.

3

▲ **Pick up first wrap.** Wrap around the shafts to take up any excess strap, but leave enough to allow you to return to the buckle on the holdback strap. Each wrap of the strap is spiraled smoothly around the shaft so the wraps stack forward. On the way back to the buckle, the first wrap is picked up so the tip of the holdback strap can pass under it. Before going through the buckle, make sure the holdback strap passes over the trace so the trace is encompassed by the holdback strap.

5

4

▲ **Trace is encompassed by holdback.** We wouldn't want to tie the trace to the shaft anywhere along its length, but the trace can be encompassed by the holdback. This way, the singletree can function to allow for shoulder motion. Encompassing the trace helps to contain any whipping motion the trace might have when the horse is trotting.

◄ **Adjust slack in breeching while in draft.** Before adjusting the slack in the breeching, the other holdback needs to be wrapped to the shaft, as shown above. Make sure your horse is standing square and isn't stretched out or standing as if on a bucket. When the final holdback is fastened, pass it through the footman's loop and lay it over the shaft. The holdback is then clamped to the shaft with a thumb, while the same hand pulls the shafts rearward and puts the vehicle into draft. The other hand is inserted into the breeching to check for fit. Horses and humans vary in size. An average-sized person hitching an average-sized horse (say, 5 feet 8 inches and approximately 15 hands, respectively) would put their hand into the breeching and use it as a spacer to determine fit. The hand is positioned so the width of the hand is lined up with the length of the horse's body. If your horse is smaller, you might leave a finger or two out of the equation. If your horse is bigger, you'd still put your hand in there, but go for a looser fit. The holdback strap is pulled through more or drawn out of the footman's loop as needed to get the right spacing when in draft. After the spacing seems right, the wrapping of the holdback and further encompassing of the trace is done as shown above. When the holdback on this second side is buckled into place, the vehicle is again put into draft as the person fits their hand into the breeching to check for final fit. This hand-width of slack is all that is needed for free movement of the horse when striding forward. If there's not enough space for striding, the horse will feel confined. Too much space gives the horse a sloppy feeling, which can agitate a sensitive horse.

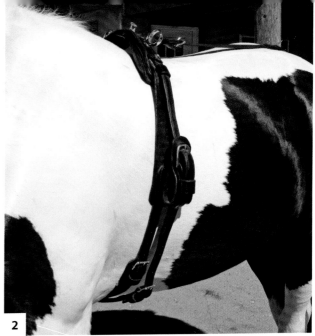

▲ **Starting the shaft wrap.** This pleasure-driving single harness comes with shaft wraps, which are also called the overgirth, shaft shackles, or shaft tiedowns. The alternative in this style of harness is carriage-type tug loops, which are easier to use but don't seem to be quite as effective for holding the shafts down. Most people using carts seem to prefer having a harness with shaft wraps because cart shafts have a greater tendency to spring up and down. Carriages (four-wheeled vehicles) have the shafts mounted to the front axle and there is never any significant upward pressure at the shaft tips. The shaft wrap is brought up from the girth area and laid over the shaft in front of the tug loop. Lay the shaft wrap on the shaft so the finished side of the leather is facing up. Remove any excess slack in the part coming from the girth. The shaft wrap attaches to the girth by a sleeve piece. The sleeve piece on the girth allows the shaft wrap to be drawn higher or lower on each side. Be sure the shaft wrap is in the middle.

▲ **Carriage-type tug loops.** Carriage-type tug loops have a tab of leather coming off the bottom of the tug loop, which goes into a buckle on the end of the overgirth. The overgirth is adjusted so the slack is removed. It is not made tighter than the girth.

▲ **Shaft wraps, further along.** The shaft wrap strap is then brought to the rear of the tug loop and you can check to see if the strap will reach the buckle. If there is still strap to use up, wrap around the shaft again to use up excess strap before going to the buckle.

◄ **Shaft wrap finished.** There's no need for the shaft wrap to be adjusted any tighter than the girth was adjusted when harnessing the horse, so remove any excess slack in the shaft wrap before going to the buckle. The horse is now attached to the vehicle. Remember that the sequence starts with the tug loops, next the adjustment of the traces, and then the holdbacks, and it finishes up with the shaft wraps.

HALTER OVER BRIDLE

It always looks much nicer and probably feels better to your horse if you try to take the halter off when your horse is wearing a bridle. Once you get the bridle on and the reins fastened into the bit, it's pretty handy to slip the halter back on over top of everything. Notice how the reins are swept to the rear by the haltering process. When you are ready to leave the hitch rail, the halter is unfastened and dropped off the end of the nose. Then your horse is ready to drive away.

▲ **A horse wearing a pleasure harness, hitched to a cart.**

▶ **Hitching a horse wearing a draft harness.** To hitch a horse to a cart with shafts while using a draft-type harness, all you need to do is form a loop in the market tug and pass the shafts through there when hitching. Heel chains on traces weren't made to hook up to pleasure-driving vehicles like this, but it is easily done by making the connection with trace extenders. If you don't need the lengthening effect of

trace extenders, you can order trace extenders that are only about 3 inches long. Holdbacks at the end of the breeching are added to the harness and wrapped and adjusted as shown above. It's a good idea to remove the lazy straps from your draft-type harness when hitching to a cart because they aren't needed and tend to get in the way. We usually leave the quarter straps, pole strap, coupler snap, and breast strap on the draft-type harness when driving single, even though they serve no purpose when used on a cart. That way, the harness is ready to go the next time we hitch a team. Most teamsters purchase a separate set of pleasure-driving reins for driving single, but it is also possible to unbuckle your coupling lines from the draft lines and use the draft lines to drive single.

TRACE SLEEVES

If you have used a draft-type harness to attach to a pleasure-driving vehicle like this cart, you might have noticed the noise. Toggle chains on the end of the draft horse traces really make a racket, especially when your horse is trotting. Trace sleeves that slip over the end of the trace and slide back over the toggle chain will fix this problem. Here, we show how to put them on and use them.

UNHITCHING A SINGLE FROM A CART

When it is time to reverse the process and get the cart off of the horse, it is equally important to follow a proper sequence to ensure safety. In summary, to unhitch, the attaching points are removed in reverse order from the way they were applied.

The last part attached was the shaft wraps, so if you are using harness that has shaft wraps, they are removed first. If you are using a draft-type or carriage-type harness, there is nothing you need to do about the tug loops to unhitch.

Next, the holdbacks are removed on both sides of the horse. In the Amish culture, with single-horse driving, the holdbacks are always left with the vehicle instead of being removed at this point. They insert a snap between the holdback and the breeching, which makes it very fast and easy to connect and disconnect the holdback from the harness. If you are using the same horse on the same vehicle all of the time, as the Amish do, having a snap there makes a lot of sense. If you order a pleasure-driving single harness from an Amish harness maker, it's not a bad idea to make sure the harness includes holdbacks. Many Amish men consider the holdbacks part of the vehicle, not the harness!

The last thing to be removed is the traces. This sequence makes a lot of sense because if the horse steps forward while unhitching, the vehicle will stay with the movement of the horse. It would scare a horse if he stepped forward a little with the traces disconnected and the shaft tips fell to the ground. It would be especially bad if your use of faulty unhitching sequence meant that the holdbacks were still connected at that point.

HITCHING A TEAM TO A POLED VEHICLE

Most teamsters seem to prefer hitching a team of drafts by driving them into position over the end of the tongue. Others will individually lead each horse into position and hook up their driving lines before doing the hitching steps shown here. When training colts, I often do a variation on that second method. The horses start out tied to a fence and the vehicle is placed so the horses are on either side of the tongue at the fence. The bridles should be on your horses and the lines attached before hooking up any harness parts to the vehicle. There should never be a moment when horses are attached to any vehicle without bridles that also have the lines attached.

◀ **Driving over the tongue.** Well-trained teams will approach the tongue and step over it from either side. As the horse steps over the tongue, the team is turned sharply to get them into position with a horse on either side. It's a nice idea to use a breeching tie when ground driving to keep the horses' bodies parallel to the line of travel.

◄ Attaching the neck yoke.

To attach the neck yoke, the horses should stand in the hitching position without needing to be held there with the lines or with people on lead ropes. Part of good training is to teach horses to be calm when standing ready to be hitched. Many teamsters grab the end of the tongue and hold the tongue and neck yoke up with their hands and legs as they fasten the neck yoke into the coupler snaps. Besides being rather difficult, this is dangerous and a good way to get slimed by the horses. As seen here, if the person hitching stays on one side at a time and reaches for the neck yoke end with their toes pointed toward the neck yoke, things are much easier.

▼ Other side of neck yoke up.

◄ **Attaching traces.** With draft horses, it is customary to attach the inside trace first, leaving a clear pathway to the side of the vehicle if the horses should start forward suddenly. Prior to hitching, the traces are attached by their heel chains (also called toggle chains) to the trace carrier hooks, which are part of the hip strap assembly on most harnesses. If you attach the trace to the carrier in the shortest possible link on the trace, the trace will be held high enough to stay out of kicking range of rear feet during fly season. Horses don't usually like the sensation of hanging a rear leg over a trace. When hooking toggle chains to singletrees, teamsters usually instruct helpers in the language of "drop." To "drop three" means to hook in the fifth link back, which leaves three links hanging down.

The main idea with hitching a team to a poled vehicle is to get the braking and steering systems of the harness attached before then attaching the traces. If you make a huge mistake and attach your traces first, your horses will be fully capable of pulling the vehicle, but they won't be able to steer or stop it. If they respond to line pressure and try to stop, the vehicle coming along behind will overtake the horses and crash into their hind ends. It's important to memorize the hitching sequence shown so that you never have to personally experience the trouble associated with that kind of hitching.

After the neck yoke is attached to the coupler snaps on both horses, it is safe to hook up the traces. Draft horse users usually attach the inside (medial) trace first when hitching. Doing so leaves a clear pathway to the side of the horses if they start moving while hitching. Pleasure-driving whips who are hitching pairs tend to lead their horses into hitching position, attach the outside trace first to contain the hind ends near the pole, then reach over the outside trace to attach the inside trace second.

UNHITCHING A TEAM

Just as you learned with single horses in shafts, the unhitching sequence is in reverse order of the hitching. That puts us at the traces first. Traces on draft teams are customarily unfastened at the outside trace first, then the inside trace. As each trace is removed, it is hung by the most forward link in the trace carrier. No matter how many people are assisting, it is the driver's job to assure that the work is done properly. If all of the traces are not removed before the neck yoke is removed and the horses are led forward, quite a wreck can occur.

Next, the neck yoke is unfastened from the coupler snaps. It is easiest if the person doing this stays to the side of the horses and lowers one side at a time. Avoid grabbing the neck yoke in the middle and trying to grapple with holding up both sides.

Many teams of horses have the bad habit of surging forward when the neck yoke is removed. If your horses act that way, it is better to make them stay put until it looks like it is the teamster's idea to leave instead of the horses' idea. If you are teaching this on a fidgety team, be aware that they might decide to back up pretty fast. Stay to the side or up on the vehicle so you don't get smashed between the horses and the vehicle.

SUSPENSION IN THE HITCH: LOOKING FOR THAT LINE

I remember the first time an attorney ever called me as an expert witness in a runaway horse case. When he told me that two people had been killed in the wreck and that the driver of the horses was saying that "horses occasionally run off for no reason," I was ready to help.

One of the first things I found out about the wreck was that it all started when the tongue fell out of the neck yoke and clattered to the street. The defendant probably didn't want to hear my opinion, but the tongue had fallen out because he hitched them too loosely, a common mistake.

After being an expert witness for several horse runaway cases, I am amazed at how many teamsters have confusing ideas about how to hitch properly. As an expert witness, I always ask the defendant, "How do you know when you are hitched properly?" Some of the interesting answers that I've heard are: "My dad said to always drop three," "You can just 'tell,'" "You do it by backing them up," and "You look at how loose the traces are." If you know a few simple things about how to adjust your horses' harness, how to adjust the toggle chains on the end of the traces, and how to secure the neck yoke to the tongue, you can avoid various wrecks. Proper and safe hitching is largely a matter of knowing where and how to look for a particular straight line. Before describing that line, let's start by defining the goal of proper hitching.

The Suspension

Many teamsters are surprised to learn that a properly hitched team is tightly suspended between the pulling apparatus and the stopping apparatus of the harness. (This suspension, of course, is only possible on vehicles and implements that have a tongue or pole. Logs and sleds on dirt are examples of things that don't need the "brakes" that the tongue provides, and therefore in such pole-less hitches the horses are not suspended.) A well-hitched team that is leaning into the collars will only have about a hand width of slack in the breeching on both horses.

▲ **Proper slack.** Be sure there's a proper amount of slack in the breeching. A hand's width of room in the breeching of each horse while in draft is all that is needed to allow free movement.

The horses are suspended tightly, but comfortably, between the pulling and braking systems of the harness. Such tight suspension assures that the middle neck yoke ring is always positioned securely back against the stop on the bottom of the tongue. In one runaway case for which I served as expert witness, the driver claimed that a bicyclist had caused her horses to rear straight in the air while they were at a dead run, which made the tongue fall out of the neck yoke. It was my opinion that not only was rearing at a dead run impossible, but such drastic movement would not cause the tongue to fall if they were properly hitched. On a well-hitched team, it doesn't matter what the horses do or the direction they go: up, down, forward, back, right, left, or any combination of the above, the neck yoke will stay safely back against the stop on the bottom of the tongue.

Teamsters whose tongue falls out of the neck yoke are hitching too loosely. Instead of having a hand's width of room in the breeching, there is significantly more slack. Besides being dangerous because of the tongue-dropping problem, such looseness tends to

◀ **Adjust the harness to the line.** Adjusting the quarter straps and/or the pole strap is the primary way of getting the harness adjusted to the line. Be sure the horse is standing square while you are adjusting.

▼ **Properly adjusted traces.** In this case, three links are dropped, and the traces are properly adjusted. Horses were led forward until their traces tightened to test for the fit.

▲ **Traces are too long.** No links have been dropped, and the traces are too long. This is the way of hitching that often causes catastrophic results. If the tongue doesn't fall out of the neck yoke, the other problem is that there is so much slack. When the horses step forward from halt, they bump into the load when they reach the end of the traces instead of gently taking it with them as they start to move. It also complicates rein handling because there is so much more rein to be handled by the teamster. Hitching too long can be very upsetting to sensitive horses.

upset horses. If it takes a second or two after the horses stop their feet for the breeching to slap into their rumps, the horses are going to get upset. Likewise, such loose hitching can cause an unnecessary jerking on the collars as the horses start a load. Any kind of sloppiness in the suspension between the pulling and braking apparatus should be an early warning that the horses are not properly hitched. If the breeching appears to be billowing out behind your horses in a

loop big enough to throw a basketball through (as they are leaning into a load), they are not properly hitched.

Of course, it is also possible to hitch horses too tightly. If you see horses with the hair worn off around their hind ends where the breeching rubs, they are probably hitched too tightly. This opposite extreme of hitching suspension doesn't cause the major safety problems that being hitched too loosely can cause, but it can affect the comfort of the horses.

Adjusting the Harness to the Line

Proper suspension of the horses when hitched starts by correctly adjusting the harness on the horses. The final test of harness fit, which should be done each time the horses are hitched, is to pull out on the coupler snap and see if its forward limit of travel matches up with the line defined by the hames. Be sure that the horse is standing square and not with his back feet near his front. Your horse also should not be standing as if he is stretched out. If you neglect this important harness adjustment, you will have a hard time hitching the horse properly. This is where differences in body length between the two horses are made equal so that the traces can be hooked at the same toggle link on both horses, no matter their height, weight, body length, disposition, and so on.

Such popular schemes as shortening the traces on the "faster" horse or lengthening the traces on the longer-bodied horse don't make a lot of sense to me. Toggle-chain adjustment is all about getting your

▲ **Traces are too short.** Seven links have been dropped here, and the traces are too short. This kind of hitching will pull the hames out of the collar every time. It also puts the horses into a bind that affects their comfort and strength.

horses properly suspended in the hitch. Longitudinal alignment problems (having a faster horse) are best addressed by applying more driver skill, not by goofing up your team's suspension. Body length differences are taken care of by the proper adjustment of the harness, as described above.

Adjusting Trace Length

After the harness is adjusted, the only remaining adjustment is to figure out which toggle chain link to fasten into. You figure this out by getting the horses into position on the tongue, getting the neck yoke fastened into the coupler snaps, and then fastening to the toggle chain links most likely to get "the line."

The definition of a well-hitched pair is one that, when properly harnessed, hitched, and in draft, viewed from the side, displays a straight line, defined by the hame, which extends through the breast strap, coupler snap, and neck yoke rings to the stop on the bottom of the tongue. As you can see from the illustrations, if the team is hitched too long (too-

▲ **The effect of short traces at the breeching.** A horse doesn't have much room for movement when you adjust your traces this way. You can't even get a finger under the breeching.

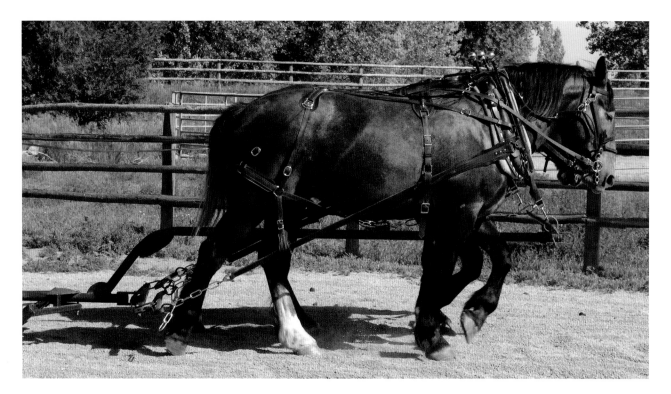

▲ **Walking with a moderate load.** This sled pulls pretty hard on sand and forces the collars back onto these colts' shoulders. Notice that the line defined by the hame extends through the breast strap and coupler snap.

◀ ▼ **Doubletree length.** If you attach doubletrees to a tongue and set them so the evener bar is perpendicular to the tongue, you will be ready to measure the correct length. Pick up a singletree as if it were in use and stretch the inside trace hook forward. The distance from inside that trace hook to the draft hole on the doubletree is the measurement of the length, not the width, of the doubletree.

loose suspension), the line defined by the hame is not straight. Such hitching allows the neck yoke ring to creep forward toward that dangerous spot for a neck yoke: off the end of the tongue.

If a team is hitched too tight, the line will be broken in the forward direction. Horses hitched this way are too tightly suspended and feel like they are being clamped on at both ends. Even when pushing into the collar, the breeching is pushing into their rumps.

Teamsters who have a hard time keeping their hames on their collars (the bottom of the hames are always pulling off) need to check that the horses are harnessed and hitched correctly. If the line is broken forward, it will cause your hames to pull off at the bottom of the collar.

Seeing the Line

A simple way to get the "view from the side" is for the driver to seek the opinion of a helpful person who is standing in the middle of a circle as the team is driven at a distance around that person. The person only needs to see the team from one side to determine if they are hitched properly. Turning around and going the other way so the person in the middle of the circle can see how it looks on the other horse is never

necessary. If both horses are harnessed properly and are wearing harnesses with traces of the same length, the line will look the same on both horses.

You can make suspension adjustment decisions on a quiet team while they are standing still as long as you have the right vehicle (something that drags or has brakes) or a helpful uphill grade. To do this, I have the driver ease off of the reins just enough so that I can lead the team forward for a step or two. When they're leaning into the load, I try to steady the horses to a stop so that they are standing still and leaning into the load. Then I can sneak around to the shoulder of one horse and get a look at how they are suspended. On some vehicles that are very lightweight, you can make hitching decisions by having the driver get off and pull the vehicle back so that the traces are tight while the team stands there. An assistant can view and report the results from the side.

How Wrecks Happen

I think that teamsters get duped into the dangerous practice of hitching too loosely (too long) because they look at the space between the horses' heels and the singletrees instead of looking up at the line defined by the hames. There are a lot of antique vehicles with tongues that were made for smaller horses. Back in the pre-tractor days, farmers used horses that were significantly smaller than the modern horses bred today. Those smaller horses were often hitched on tongues that were 8 feet, 6 inches long (effective length). Trying to hitch a long-bodied modern pair of horses on a wagon tongue that is only 8 feet, 6 inches long will not work! When teamsters try to make it work by lengthening their traces (hitching out at the end of the traces), the suspension becomes loose and tongues begin to fall out of neck yokes.

The length of tongue I prefer is about 9 feet, 6 inches long (effective length). If the tongue on your wagon is too short, there are a couple of easy fixes to try before resorting to replacing the tongue with one that is long enough. Sometimes there's room to move the doubletree draft hole closer toward the vehicle to create more length. It also might be possible to move the neck yoke stop forward at the end of the tongue by adding a steel extension. Unfortunately, as every good carpenter knows, you "measure twice and cut once" to avoid coming up short. When it is determined that a tongue is too short, the best answer is to purchase more materials and make it long enough.

Figuring Tongue Length

In order to know how long to make the tongue, start by measuring the actual length of the traces on the harness (on most draft-horse harness, that'll be 8 feet or 8 feet, 2 inches). Then you need to measure the effective length of the doubletrees—not their width, which is more often measured.

To measure the length of doubletrees, hang them in position on a tongue. Position them so that they are perfectly perpendicular to the tongue, and then hold up a singletree so you can measure the distance from the draft hole on the doubletrees to the inside of a trace hook on a singletree. Most doubletrees are about 1 foot in length.

The length of the traces plus the length of the doubletrees add up to a good length for the tongue. This set-up will have the horses hitched so the singletrees are fastened to about the middle of the toggle chains. Since modern-day draft horses and mules are getting taller and longer with each breeding generation, I suggest adding about 6 inches of extra

length to the equation to make the effective tongue length (for horses in 8-foot-long traces) 9 feet, 6 inches. Then the horses will use almost the entire length of the trace when suspended in the hitch.

For hitching medium- and large-sized ponies and saddle horses, I prefer that the harness be made with 8-foot-long traces so that I can conveniently hitch onto most equipment and vehicles. Extra space between the heels of the horses and the singletrees is not a problem. If you are planning to hitch smaller animals, the starting point for trace and tongue length considerations is to measure the horse's body for a minimum safe trace length that would give plenty of clearance for a low singletree. Add on 18 inches of length, which would be a good overall trace length. Then add up that trace length plus the doubletree length (throw in a few inches to put you at the end of the traces for hooking) to figure out the effective length of tongue to use.

FASTENING THE NECK YOKE

One additional technique to add even more safety to hitching is to always be sure that the neck yoke is fastened to the end of the tongue. The way most teamsters dismiss this important safety precaution, you'd think that having a slip-off neck yoke was some sort of a safety device. Slip-off neck yokes are definitely a safety liability! The only reason neck yokes were made to easily slip on and off was for the convenience of the farmer who would only use one neck yoke for his several vehicles and implements.

Even if you are properly hitched, having a trace come unfastened from a singletree can allow the horses to get far enough forward on the tongue to have the tongue drop out of the neck yoke. If the end of the tongue doesn't have a safety catch to contain the yoke, I chain the neck yoke to the stop.

Of all the stupid things people do to get themselves and others hurt while driving a team, I think the most common is hitching up without really knowing how. When a team is not properly hitched, the wreck caused by that lack of knowledge will usually be catastrophic. Full pulling ability without any of the stopping or steering ability (when the tongue drops) is a very bad thing. Safe hitching is largely a matter of looking for that line while harnessing and hitching.

Chapter Five

Lateral Alignment

We were still 100 feet from the top of the chairlift on the bunny hill, and the crowd was starting to gather. This was to be our fourth and final attempt on the beginner's slope for the day. What a frustrating day it had been! Although there were other 30-somethings on the bunny hill that day, most of them were there as parents, expertly teaching their squat, sturdy little children (upright as penguins) who were one-tenth of my age.

For the most part, I could tolerate being the biggest kid on the bunny hill, with my beautiful and patient wife as my instructor.

Really, the only part of this new experience that I definitely didn't like was the process of getting off of the two-man chair at the top of the lift. It was there that I was having alignment problems with those two boards they'd outfitted me with at the ski rental shop.

Many attempts and failures later, I was probably the person most strongly convinced in the world of the idea that alignment problems are not fun—unless you are an onlooker. Although I've cooled off somewhat in my conviction about that idea since I've gotten better at skiing, I still hold the idea as one of life's great truths.

One pursuit that, like skiing, relies upon good alignment for success is team (or pair) driving. I've often seen teams where one horse goes ahead of the other continually, one or both horses travel with their heads to the inside or to the outside, the horses stagger together or apart when starting a heavy load, the horses spread apart when going up a hill then come together when going down, or vice versa. If your team exhibits any or all of the above symptoms, you have an alignment problem that needs to be corrected. If a horse's body is being bent as the team is going straight forward, or if a horse must stagger for balance while working, it will have a hard time being happy, calm, and trusting. It is no coincidence that in poorly aligned teams, one or both of the horses often act nervous.

There are three dimensions to the alignment of a team.

• **Vertical Alignment (Height).** This is the dimension that is often most worried about and costs the most to achieve, but it matters for the comfort and efficiency of the horses. Vertical alignment adjustments are mostly for cosmetics—putting the taller horse on the low side of the road, shortening the breast strap on the shorter horse, etc.

• **Longitudinal Alignment.** This is how well the team lines up nose to nose, which is best seen from above. If a teamster keeps his or her horses parallel to each other and keeps the doubletrees perpendicular to the tongue at all times, he or she has perfect longitudinal alignment. One horse being ahead of the other by an inch or a yard is a longitudinal alignment problem. The problem is that the trailing horse is in the slack, not making contact with the bit, when this happens. Longitudinal alignment is difficult for the amateur teamster to achieve and happens without apparent effort by the experienced teamster. Much more will be said about longitudinal alignment, but the focus of this chapter is lateral alignment.

• **Lateral Alignment.** This has to do with how far apart the horses are side to side and the factors that make them parallel to the tongue at that distance apart. What is it that determines the spacing goal between two horses? In row crop work, it is obvious that the distance between pathways among planted rows is the primary consideration. But when hooked to a wagon going down the road, how far apart should the horses be, and how do you make that adjustment? I've seen teamsters going down the road with the apparent lateral alignment goal of having a horse on either side of an 8-foot-wide roadway (with the front parts of their bodies, at least). Inversely, I've seen teams where it appeared that the alignment goal was to have them both occupy the same space between the horses.

There are four elements that determine how far apart a team travels:

1) Width of Available Roadway. No matter the effects of determiners of width described below, if you cross a narrow bridge, go through a narrow gate, or travel down a packed sled trail with belly-deep snow on each side, these pathway constraints tend to push a team closer together.

2) Line (Rein) Adjustments. Lines determine where the bits are centered; it is an ancient and true axiom of horse behavior that horses seek the center of the bit. How far forward or back the coupling lines (also called check lines, crossover lines, stub lines, splice lines, brace lines, etc.) are adjusted on the draft line, and the presence or absence of line spreaders (and, if present, their length) are how the lines are adjusted. Hame rings and line spreaders control team spread (distance apart) by controlling the amount of deflection in the coupling line. Coupling-line buckles control spread by length adjustment. Moving coupling lines farther forward on the draft line spreads the team apart, and moving the coupling lines back brings the team closer together.

3) Neck Yoke Width. Neck yokes don't affect alignment much when going uphill, but put 3,000 pounds of sled and people on the breechings of two horses going down a steep hill and you have a major determiner of alignment.

4) Doubletree Width. Going downhill on a wagon, the doubletrees don't affect alignment at all. However, the same doubletrees on a wagon in deep mud, going up a hill with a full load of 75-pound bales on board, will be a major alignment factor.

The horses' lateral alignment goal is affected by these four different factors. Although they might not operate all at once, they all have to be in agreement.

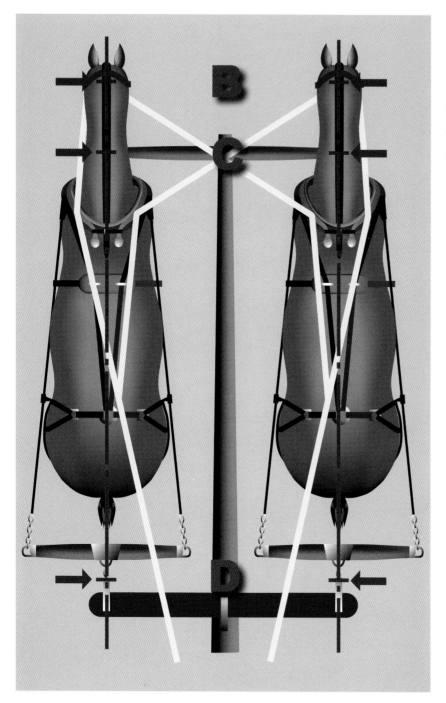

◀ **Ideal lateral alignment.**
In teams or pairs that show good lateral alignment, you could say that "B = C = D": bit centers equal coupler snap centers equal doubletree width. Any deviation from this equation will twist the bodies of the horses to some degree and make the horses uncomfortable with their work.

Each horse should travel (when going straight ahead) with its body making a straight line from the center of its nose to the center of its tail. Each should be parallel to the tongue and equal distance from it, no matter the load or terrain.

If the lines are adjusted so that the centers of the bits are 56 inches apart, and the team is pulling a heavy load on a 42-inch doubletree, the team's bodies will be making an open V shape as they go down the road. With enough use where the lines are

▶ **A lateral alignment problem.** If the lines are adjusted so they are set wider than the doubletrees and neck yoke, the team will move forward with an open vee look to their bodies.

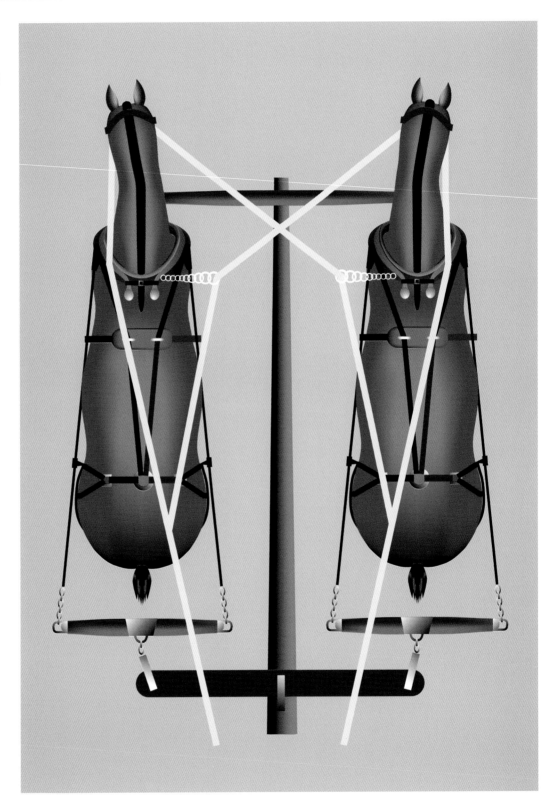

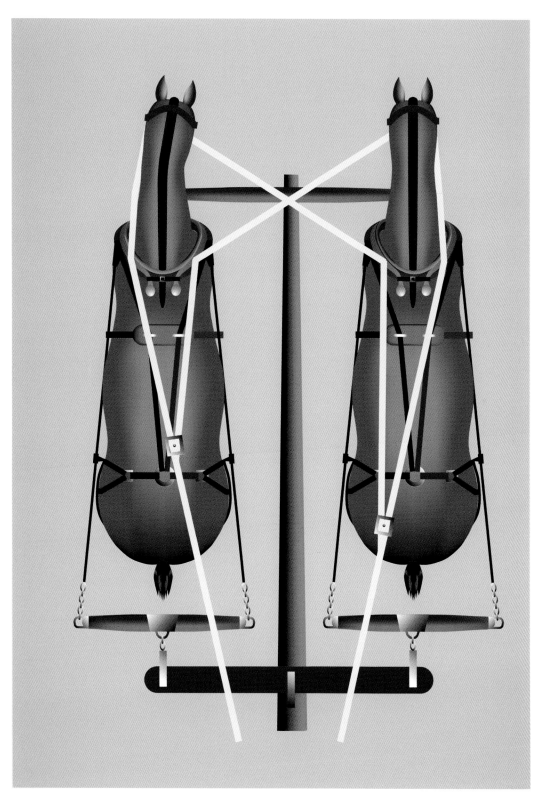

◄ **Maladjusted lines.** When one line is adjusted and not the other, heads are tilted.

set wider than the doubletrees, one or both of the horses will learn to stay centered on the doubletrees from their tail to their collar and bend their neck outward from the collar forward. This way, part of their body is made to accommodate the doubletrees and part accommodates the bit. It is not a comfortable arrangement for the horses, and it is commonly seen.

I've noticed that among many draft-horse people, when there is some lateral or longitudinal alignment problem, the first adjustment that is leapt upon is to asymmetrically adjust the lines (shorten or lengthen one coupling line, but not the other). In cases like the one above, where the lines are set wider than the doubletrees, the more flexible, accommodating, and better-trained horse will be the first to realize that he has to travel with his head out. He is rewarded for being the better horse by having one of his lines adjusted so that his bit is henceforth tilted crossways to his direction of travel. To further accommodate this added alignment insult, he learns to move closer to the other horse in order to get enough slack to get his bit straight with the direction of travel. It would be much easier for the horse if the teamster had symmetrically adjusted both lines to match the doubletrees and neck yoke.

Lines are never to be adjusted asymmetrically (where just one coupling line is moved). A horse going with his head in or out (assuming everything is adjusted properly) is exhibiting a serious training problem that is corrected only by patient retraining of bit response and rebuilding confidence with load pulling. Pleasure pair-driving reins are manufactured with a definite right and left line being different from each other to allow the teamster to drive from a position directly behind the right-hand horse. Most draft-horse lines are identical to each other because the driver is expected to take up a position between the horses.

THE LEANING TEAM

A few years ago, I got a call from a sleigh ride outfit in ski country that had a serious lateral alignment problem with a pair of mares. Their problem, which they wanted me to fix, was pretty amazing. When these mares were started off on the sleigh, they immediately leaned into each other, crushing themselves together on a horizontal line down their sides that was almost up to their backbones. To lean into each other that much, they had to throw their feet way out to the side, and dig into the snow with the sides of their hooves. Standing still (before the leaning started), they occupied about 6 feet of road width. Under way, they needed at least 10 feet of roadway because of their sprawling legs. While they had perfect vertical alignment at all times, it was interesting to watch their vertical height drop by about 2 feet when they started going forward.

This sleigh ride team was being driven on 38-inch doubletrees. The lower cost of shorter doubletrees was the apparent logic. On the more modern, thicker horses that people drive now, the more narrow 38-inch doubletrees have the horses so close together that they are bumping into each other. Unfortunately, it is normal for horses to push back when they are leaned on. For wagon work, I like 42-inch doubletrees a lot better.

The sleigh ride outfit used a 42-inch neck yoke on the leaning mares. The effective width (end ring to end ring) of the neck yoke should equal the effective width of the doubletrees (center of singletree clevis hole to center of singletree clevis hole). When I removed the lines to check their adjustment, the draft line was equal in length to the coupling line when measured from the coupling line adjusting buckle forward. Just as a ballpark figure, I've found that average-sized draft horses (with average neck thickness) will center their bits on 42 inches (without line spreaders), if you adjust the lines so that the coupling lines are approximately 8 inches longer than the draft line. In theory, the leaning mares were being driven with their bit centers 42 inches − (2×8) = 26 inches apart.

There were no line spreaders on the leaning mares. If the lines had been left the way they were (no difference in length of coupling line and draft line), but 8-inch-long line spreaders used (these effectively extend the position of the upper hame ring 8 inches toward the center on each horse), the centers of the bits would have been at 26 inches (before spreaders) + 16 inches (spreaders) = 42 inches.

Exactly calculating any line adjustment results on paper would be enough to baffle a rocket scientist.

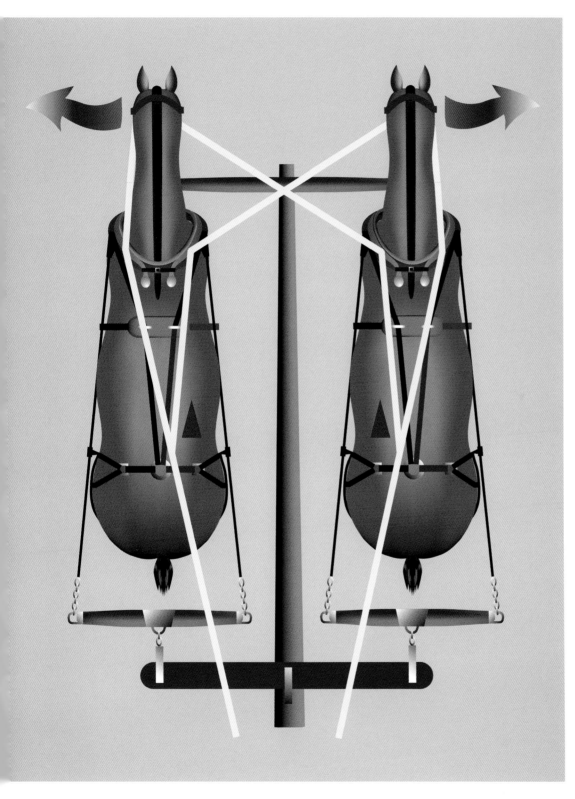

◀ **Adjusting the coupling lines forward.** By moving the coupling lines forward on the draft lines an equal distance, the team is widened out. Each inch of forward movement on the lines makes an inch of spread between the horses' bits. If both coupling lines are moved forward 8 inches, the overall effect would be to widen the bit centers by 16 inches.

▶ **Adjusting the coupling lines rearward.** When the coupling line buckles are moved to the rear by the same amount on each side, the horses are brought together. An inch to the rear on both coupling lines brings the team 2 inches closer at the bits.

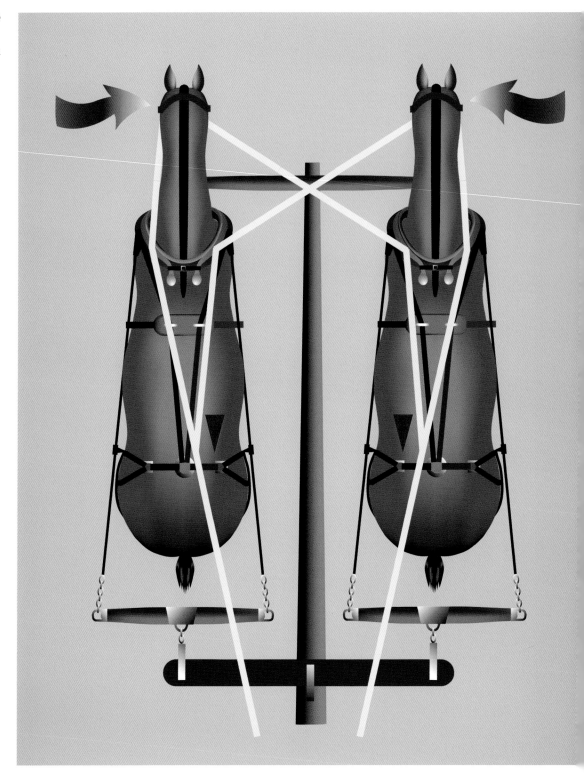

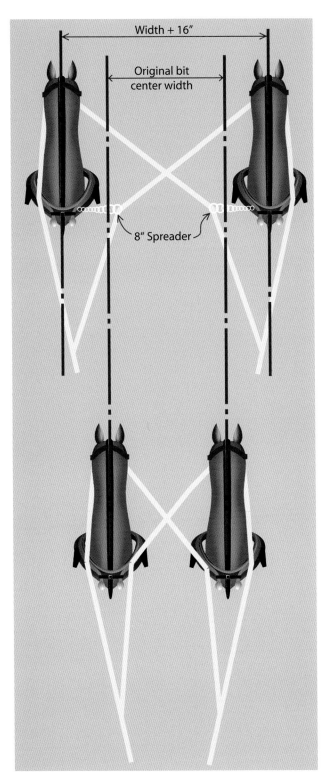

Width + 16"

Original bit
center width

8" Spreader

◀ **The effect of line spreaders.** Line spreaders attached to the hames serve to widen out the distance between the horses by removing some of the deflection of the coupling line. Each inch of spreader length creates an inch of spread between the horses. Using a pair of 8-inch spreaders would move the team apart by 16 inches if no other elements were changed.

Fortunately, there is a better way to see if your adjustments will work. Set up a situation in which the position of the bit centers is the only thing affecting lateral alignment. This means ground driving (not hitched) on open ground (no predetermined pathways) with the horses on the bits (longitudinally aligned and no slack in the lines). In this situation, you can get a good idea of how the bits are centered by measuring the distance between the center of their noses, their neck yoke coupler snaps, or the center of their tails. The neck yoke that will be used when hitched is a good measuring stick. You'll have to do some walking around to be sure they're really on the bit. You'll also have a hard time measuring anything while walking, so measure their center after you stop and make sure they don't change their distance apart as they go from walking to a stopped position.

To correct the leaning mares' problem, I first adjusted their lines to give a little distance between them. I thought 42 inches seemed more appropriate than 26 inches for this fat pair of Belgian mares, so I moved the coupling lines forward 8 inches on each side. Not surprisingly, when ground-driven on the new adjustment, they didn't separate as they should, but continued to lean together. They were more reliant on their extensive prior training than on the effect of the new bit positions.

Undaunted, I requested that the 38-inch doubletrees be exchanged for 42 inches, and the mares were hitched. I reasoned that by having all alignment affecters in agreement at 42 inches, they would automatically cease their crushing. Not so.

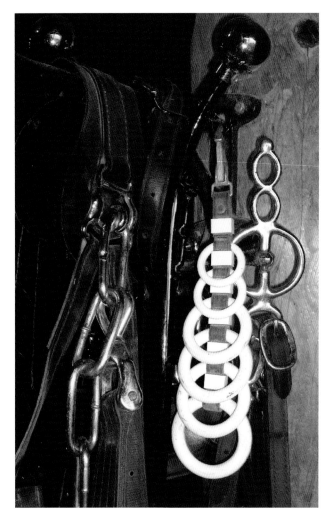

▲ Line spreader on a hame. Line spreaders snap into the little ring near the top of a draft hame. The coupling line passes through the end ring of the spreader instead of through the upper hame ring.

Even going up a steep hill with a heavy load where the traces were rigidly forcing their bodies apart, they clung to their ways. I was amazed at what training, good or bad, will do to a horse.

Thinking hard now, it occurred to me that the right-hand horse wanted to lean left, and the left-hand horse wanted to lean right. It would serve my purpose perfectly to switch sides on them, but I held back from that decision for a moment. When

switched, would they both fall down without the other to lean on? Finally, with the powerful reasoning that they weren't worth much the way they were, I thought, "So what did it really matter if switching them caused them to fall down?" I went ahead and, with great trepidation, switched them. They didn't fall down, as I had feared. Instead they drove off like a perfect pair of horses and have been behaving that way for over four years now. It is odd to find a situation such as the one with those mares in which none of the determiners of lateral alignment were in agreement. If just one of the alignment points is out of agreement, it is enough to cause major problems. Also, it's odd to see an alignment problem in which one of the major causes is that the bit centers are set narrower than the doubletrees. By far, the most common alignment problem I've seen teamsters inflicting on their horses are bit centers that are wider than the doubletrees.

If you get everything lined up perfectly and one horse or both horses still stagger into or away from each other when starting a load, or they travel with their heads out of line, what then? Just as with this team, previous training sometimes overrules correct alignment. Give them some time to learn that the contorted body positions and staggering for balance is no longer needed. As a horse becomes more comfortable because of proper hitching, it will relax and behave. Your problem may also be sore mouths. Have your vet look for wolf teeth and sharp hooks on molars, which are major causes of undesirable head positioning.

As we rode the chair to the top of the lift line that awful day, I finally summoned the skill to get those skis parallel to each other, the right distance apart, and far enough from my wife's skis that she couldn't help. We sailed down that icy mound with unbelievable ease. As we let out victorious whoops of joy, I looked around with surprise at the faces of all those who'd hung around to see me fall again. They were all smiling back at us with joyous looks on their faces. Although alignment problems are not fun, overcoming alignment problems is wonderful. I wish you the best of luck at fixing yours.

Chapter Six

Rein Grips for Singles and Teams

Whether your use of horses for work will be a positive pursuit or just one of those interests that never amount to much might depend on something that seems like a little thing. That "little thing" is how you hold the lines (reins). Effective horsemen and women seem to have a knack for being in the right place at precisely the right moment with their rein pressure. A little bit of sloppiness, ineptitude, or weakness in rein handling could turn something that should have been a non-event into a very bad episode.

There has been a lot studied and written on the subject of rein handling by those who ride horses. I don't think there is a reason to assume a style of rein grip that works for riding race horses and dressage horses and jumping fences wouldn't work for driving, too. The English grip is one of three grips shown in this section, but it is probably the most appropriate grip for most driving situations. One of the best things about the English grip is its strength. By opening the English grip a little and looking at the pathway of the rein through the hand, you can see how it gains its strength by the double L bend it puts in the rein. Another great thing about the English grip is the way it can be moved forward on the reins in a quick and strong manner, keeping rein tension even to both reins and constant as the forward movement is carried out.

Going up hills pushes the mass of the horse out through the opening of the collar or over the breast strap. The horse also tends to extend his head and neck as he pushes into a load while ascending a hill, pulling the reins forward in the driver's hands. Going downhill, the horse's body mass compresses as the feet gather together to hold the vehicle back and the horse's mass pushes over the breeching. Holding back on a steep hill with a poor grip could cause slack to appear in the reins as the horse comes back toward the driver.

While driving, the teamster or whip must be able to handle about 2 feet of rein under ideal conditions. Poor hitching complicates good driving because the length of rein along which the driver operates his hands usually becomes greater than what is humanly possible.

The Achenbach grip is another classic rein grip that is very popular among pleasure-driving whips around the world. It is different than the other grips shown because this grip allows for a free hand to operate a whip. Pleasure-driving whips are definitely required to carry a whip in hand as a safety measure to keep the horses moving forward. I don't often use a whip on my horses because I have them trained to go forward on rein and audible cue. I teach people to drive, usually on my own horses. The last thing I want to do is attempt teaching a person who is already overwhelmed with learning to handle the reins to also use a whip on my horses. The biggest problem isn't what I'm afraid they might do, or the training they might undo using the whip on my horses. The biggest problem is that novice drivers often end up bashing the instructor in the face with the whip—always by accident, but it still hurts the same!

The third grip shown is what I call the comfort grip. I'll be the first to admit that if the lines are wide and thick and the day is long, you'll see me using the comfort grip because it is easy on the hands. I won't use it when things are happening, but when everything is going along just fine on slack reins, I might use it. There are a few problems with the comfort grip. For one thing, it is a weak grip. The rein passes through the comfort grip without the slightest bend. The grip strength of the driver is the only thing that makes it work. Another thing I don't like about the comfort grip is that it doesn't shorten quickly, evenly, or with strength. Most folks who practice the comfort grip use the crab technique of rein shortening, executed by a pitiful weak shuffling of the hand on a slack rein.

English grip

▲ **English grip:** This is the grip used around the world where high-performance horses are ridden with both hands.

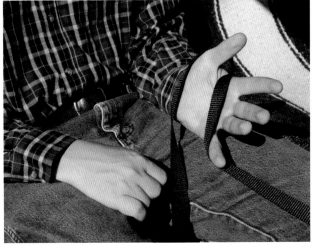

▲ **Open English grip:** The double "L" bend of the rein when it is lying in the hand is what gives this grip its great strength.

Achenbach grip

▲ **Achenbach rein handling:** The left hand is the anchor hand, the one that maintains driving duties while the other hand carries the whip. The left rein is the one on top and the right rein is placed between the second and third fingers.

▲ **The right hand assists the left hand, as needed:** When not needed for assisting the left, the right hand returns to holding the whip. The right hand reaches onto the reins with a special splitting of the fingers shown here. The second and third finger stay together as the index and pinky finger do the splits. Left and right reins are inserted into the right hand at these split spaces.

▲ **Left turn with assist:** For strength on the reins during a left turn, the right hand is placed on the left rein so the index finger of the right hand grips the left rein.

▲ **Right turn with assist:** For strength on the right rein, the right hand can assist. The right hand grips the right rein with the ring and middle finger.

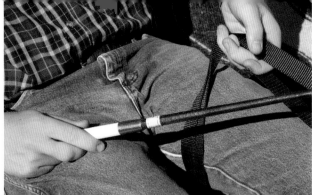

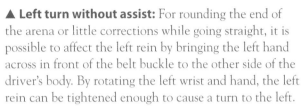

▲ **Left turn without assist:** For rounding the end of the arena or little corrections while going straight, it is possible to affect the left rein by bringing the left hand across in front of the belt buckle to the other side of the driver's body. By rotating the left wrist and hand, the left rein can be tightened enough to cause a turn to the left.

◀ **Right turn without assist:** To turn gently to the right, often all that is needed is for the driver to rotate the hand and wrist forward while curling the hand under. This tightens up the right rein for mild corrections and easy turns without needing to use the assisting hand.

▶ **Stopping and backing up with two hands:** To pressure both reins with great strength, the assisting hand is inserted into the reins where it gets a grip. Then both anchor and assisting hands pull to the rear. The anchor hand lifts up as the assisting hand pulls.

Comfort grip

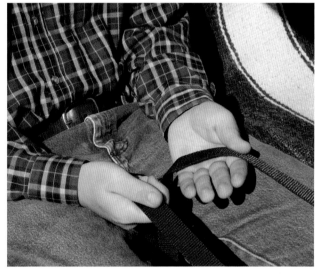

▲ **Comfort grip:** This grip is comfortable, but it's weak. Those who use this grip say it somehow makes them more empathetic with a horse's mouth. Empathy is in the brain, not the hands. Maximized potential for strength and effectiveness is what you want in your hands.

▲ **Open comfort grip:** The hand is open to show how this grip doesn't bend the rein.

◄ **The crab technique with comfort grip:** With a slack rein, it is possible to inch your hands forward at a glacial rate. Slow and slack are the problems with using the crab technique to shorten the reins.

Rein shortening

▲ **Full cross technique of rein shortening:** While one hand is reaching forward for a new forward grip on the reins, the other hand has the reins crossed in it. This grip maintains the driver's ability to steer right or left, and pull or release the reins for stopping or urging forward. Tilting the hand toward index finger or pinky finger causes heads to go from side to side. Pulling to the rear with the hand and arm causes stopping or rating.

▶ **Reaching forward with the other hand:** While a full cross is in one hand, the horses are effectively driven as the free hand reaches forward for a new grip.

▲ **Open full cross:** If you look at what is in the driver's hand, you can see the reins are fully crossed over the palm.

▲ **Forward full cross:** Another full cross is put onto the reins by the forward hand. This full cross also allows the reins to be handled with one hand, and it can all be done even with highly pressured reins.

▲ **Back to rein in each hand:** The rear full cross is released as the forward hand becomes active, then the hands can be returned to a rein in each hand grip.

To be an effective horseman or woman, you need to learn how to shorten reins while the rein is being used (pressured). The crab technique will only shorten a rein by about an inch every 10 seconds, which is not enough to be effective. The comfort grip is not a grip you'd ever see a rider use. Riders who ride with the comfort grip are called dudes.

GRIPS FOR TANDEM, UNICORN, AND FOUR-UP

Tandem, unicorn, and four-up all drive with the same types of grips. The three different grips we teach for handling a four-up are full cross, English, and Achenbach. These aren't the entire range of grip possibilities for this purpose; they're just the ones that we like the best.

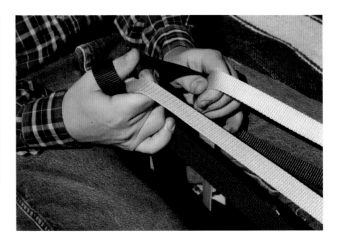

▲ **Full cross:** The reins for the leaders (the pair on the front) come in at the top of the hand and exit out the bottom. Wheelers' reins come in at the bottom of the hand and exit out the top of the hand. The two reins cross over each other in the palm of the hand. Rotating the wrists and hands in a curl toward the top of the hands activates the leaders' reins only. Rotating the wrists and the lower part of the hand activates pressure to the wheelers' reins.

▲ **How to make small adjustments.** Small adjustments to increase tension in one of the reins can be made by grabbing the rein with the opposite hand and pulling it through your hand from behind. Small adjustments to relieve unwanted tension in a rein are made by loosening the grip that is holding the rein in place so that it slides out.

▲ **How to make large adjustments.** To move the hands forward a great distance, the reins are crossed into one hand while the other hand reaches up to put a full cross in further forward. Then the hands return to their original grip further forward on the reins.

▲ **English grip:** The more horses you add into the English grip, the more the reins stack up in the hands. Wheelers are below the pinky fingers. Leaders are between pinky and ring fingers.

The full cross grip is definitely the strongest and the easiest way to drive a four-up when you are learning to drive this many horses. It is straightforward and easy to understand because the reins are separated by the width of your hand. It is also the most powerful grip for driving these multiple hitches, making it ideal for driving under difficult circumstances.

The English grip should look pretty familiar to you, thanks to its use in the previous section. We sometimes also call this grip the American teamster because it seems to be pretty popular with the big-hitch drivers in the United States. This is the grip that lends itself best to adding on even more horses. Just keep stacking pairs of lines into the spaces between your fingers, with the leaders on top and the wheelers on the bottom and you can drive eight-up at the same time.

The Achenbach grip is presented here because it is a classic seen all over the world where these kinds of multiple hitches are driven. As with singles and pairs, the main benefit of learning Achenbach rein handling is that it is designed for carrying a whip. Achenbach rein handling is definitely the most difficult method of rein handling out there. Some people spend years taking lessons from experts in Achenbach rein handling before they begin to feel somewhat proficient. It is very impressive to watch an Achenbach rein handling pro handle a four-up through a difficult course.

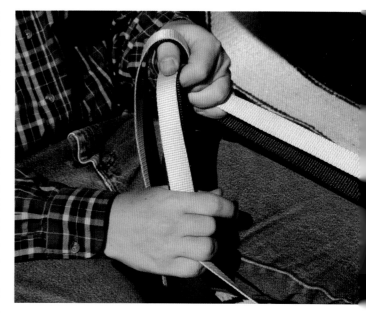

▲ **Full cross to change grip:** Put your full cross into one hand while leaving your reins laced into the hand that is sliding forward. The hand that moved forward is dragged to the rear. Meanwhile, the teamster feels for the desired tension in the reins of that hand as the hand comes to the rear.

▲ **Achenbach grip for four-up:** As with the other two grips, the leaders are toward the top of the hand and the wheelers are toward the bottom of the hand. The leaders are split between the top and bottom of the left index finger, with the left rein on top as it was with singles and pairs. The wheelers are split between the top and bottom of the middle finger. The interesting thing about this grip is the way the right leader rein and the left wheeler rein are in contact with each other in the anchor hand. That sort of arrangement makes it possible to give lessons on this subject for years and still have a student that needs to learn more! I've been told that once the whip gets the reins all adjusted for going straight with the anchor hand, they are to keep a grip on that setting until the end of the drive.

▲ **Slide area:** There's a lot going on in this rein gripping area between the fingers. With this style of driving, the reins are either gripped or slipped between the fingers as needed to get the tension on the reins that the driver is seeking. Going around the turn at the end of the arena might require the hand to be slipped to the rear a little so that one of the reins can be tightened for the turn. As the team nears the end of the turn, a cross is taken and that hand will slide forward on the reins; it can then be brought back until the tension is reset for going straight. It takes a lot of getting used to, but it happens without a thought for experienced teamsters who can make this look easy.

THE BICYCLE DRIVER TRAINING DEVICE

My friend Larry Helburg and I were out to breakfast one morning to celebrate the completion of yet another two-day workshop. Over coffee, Larry began to insist that we invent a driver-training device. He was tired of seeing rank beginners handling the reins on our single horses and teams. So many people who are learning to handle the reins on horses have a serious lack of skill. A habit of too much steering-wheel handling seems to complicate the learning of rein handling.

We started out thinking about an idea I'd been working on for a few years that involved harnessing up a pair of tall sawhorses. Then the sawhorses could be moved around and the coupling lines could be adjusted in various ways to illustrate how coupling lines function. I thought it was an okay idea, but Larry wanted something better. He asked if I thought that a pair of bicycles could be used instead of the sawhorses. I remember answering that I thought the idea would work just fine if we could figure out how to make the bicycles stand up. Larry thought he could figure out how to make that happen. We immediately set out for our local bicycle junkyard, where I purchased a pair of reasonably matched old bikes. Larry took them home to his shop and worked on all of the balance considerations and issues involved with joining the two bikes together in a way that simulated a team of horses.

At that point, we hadn't figured out that our new pair of bikes could be pushed. For the first year or so that we had the bikes, we hauled them around the country wherever we were teaching. One of us led the pair of bikes with a lead rope fastened to the lateral

◀ **Bicycle training device as a pair.** We rigged a couple of bicycles to use as a driver training device. For those who have too much steering wheel experience and not enough rein handling, this set-up will get them going on a steering system that is definitely horsey. Handling reins for right and left turning is quite a bit different from handling a steering wheel.

▼ **Bicycle training device as a four-up.** Then we got a few more, and hitched them together as a four-up. They drive just like horses except they have this maddening habit of simply going where you steer them. They never help you out like a horse will do.

joining bar. Then the person learning to drive would steer the bikes from behind. People seemed to be pretty impressed with the concept, and it did a nice job of showing how team lines function.

One day it finally hit me that the bikes should be pushed ahead of the driver by the driver's own forward motion. Immediately on the heels of that idea came the idea that if it would work for a pair of bikes, it would also work for a four-up and possibly even more. Larry went back to the shop to design the push poles and I found another pair of used bikes so he could build a four-up. With the first drive of the new set-up, we knew we really had something this time!

Soon after the four-up success, I got the idea of putting a steering tongue on the front of a golf cart so the golf cart pushes the bikes ahead as the driver steers the golf cart with the reins, just like a team of horses. Next we tried to drive our first six-up of bikes on the front of the golf cart. My son Nate and I tried it for a few days and decided that it was definitely beyond the capability of a human to handle that many bikes with reins. We gave up on the idea and went back to driving four.

As some of the draft-horse hitch drivers discovered our invention, they immediately fell in love with it and wanted to drive six-up with the bikes. It didn't take us long to see that these hitch drivers definitely could drive a six of bikes. We went back to six-up ourselves, and there it was. Nate can drive a six-up of bikes like a wild man. He can make those bikes fly around our driveway with the pedal to the metal! The latest development with the bikes is eight-up. Several of the hitch drivers in the country have latched onto the idea. There was even an eight-up bicycle hitch-driving competition at a major draft-horse show in Iowa.

The thing we like about the bicycle driver training device is that it does what we set out to accomplish with it. It allows a driver to gain lots of experience at handling the reins without involving horses in the learning process. This experience happens in a perfectly safe way with the bikes at relatively low cost to the learner. If you had to wait until you owned an eight-up of well-trained horses to learn how to drive an eight-up, that might take a while. Once the bikes are set up, they can be left parked in a garage so you can take them for a drive whenever you feel like honing your rein-handling skills.

The hitch drivers who own our bikes say that bike driving definitely makes them better drivers. The bikes are a bit more difficult to drive than horses because the bikes don't trace steer at all. (Trace steering is where the traces on either side of the horse override whatever the horse is feeling on the bit. Unskilled, beginner hitch drivers tend to concentrate on driving their leaders and count on trace steering to handle the rest.) Being more difficult than the real thing is exactly what you want in a good simulator.

DRIVING TOWARD LIGHTNESS

A few weeks ago I went to the big draft-horse sale in Waverly, Iowa, to buy a new team of horses for a sleigh ride business in Colorado. Like many buyers of "using horses," they wanted a big, well-matched, beautiful pair of horses. They also wanted them to be quiet, well-mannered, and easy to handle. Out of the many teams of horses at the sale, there were only a few that fit the order. There were plenty of big, well-matched teams, but there weren't many pairs that looked like they'd be safe for a relatively inexperienced driver to use for giving sleigh rides. The most common flaw that made most teams unsuitable was the way the horses moved forward with hard pressure on the lines.

When a horse needs to be restrained to prevent him from breaking into a trot, he'll usually display a characteristic physical appearance. Riding and driving horses that are leaning on the bit usually move with their heads elevated. These are closely associated behaviors. Another characteristic of the "elevated head" and "leaning on the bit" syndrome is the horse's demeanor. Instead of being calm and predictable, they tend to be highly reactive to things that are happening around them.

Traveling Line Pressure is Not Needed

The most common obstacle in preventing excess pressure is the belief that driving and riding horses should travel forward with pressure on the lines. This belief quickly results in a horse that quits thinking and becomes good at pushing.

Tuning out line pressure is simply a survival technique. It's the only way the horse can do his job.

▲ **Let the lines go slack while walking.** The lines don't have to be hanging loosely to get good results, but there needs to be no pressure when the horses are acting right. When doing field work like mowing, it's nice to be able to loosen the lines when your team looks like this.

There is probably no bigger turnoff to prospective drivers than seeing how most driving and draft horses can't go forward without significant pressure on the lines. While most drivers seem to think that this is some sort of a safety feature, many are beginning to see it as the opposite.

When horses are taught to advance with pressure, the amount needed to keep them in the selected gait and amount needed to stop them is definitely going to increase over time. If you need to hold pressure to keep your horse walking, it's going to numb him progressively to bit pressure. Such driving is not what

I'd call driving. It is more like waterskiing. It is very common to see horses that require 50 or more pounds of pressure just to keep them in gait. Stopping those types of horses usually requires much more pressure because the horse's ability to detect the change from "don't leave" to "stop" has been erased by the constant pulling.

Getting Your Horse Not to Pressure the Bit

Pressuring the bit is correlated to how horses pressure their handler while being led. Soft-on-the-line horses usually lead by following their handler instead of dragging him around on the end of the lead rope. Some of the most significant training for being light on the lines can happen while the horse is wearing a halter and lead rope.

Whether a horse leans on the bit or is light on the lines is ultimately determined by how the driver drives, so let's focus on how to handle the lines. Being hard on the lines is what comes naturally between horses and their driver. To prevent that from happening, the driver needs to operate with a set of principles that is different from most. The most important principle is remembering to stay relational. If you start some communication with the lines, be sure that the horse responds before you remove the pressure. Unyielding pressure only results in a breakdown of communication.

When the horse learns to go forward without continuous line pressure, he's also learning the liberating concept that being left alone (going on slack lines) means that he is doing the right thing. If the horse speeds up, his peaceful state of being left alone is interfered with as the bit is pressured. Most horses figure it out in a hurry and seem to appreciate the change of principle. They show their appreciation for the new way of doing things by lowering their heads and losing that pushy, alarmed look. Now, lack of pressure is clearly perceived as meaning, "Yes! You're doing the right thing!" However, pressure on the lines will indicate that a change is required. Previously, pressure on the lines was just something to tolerate, not something to indicate a deliberate change.

When the Driver Becomes Like a Post

Stopping and rating speed are two-handed line handling operations. The way most people, both riders and drivers, are taught to stop their horses is a major contributing factor to creating so many bit-pushing equines. Most horse people are taught to stop a traveling horse by first saying "whoa," then pulling on the lines as necessary until he stops his feet. Once stopped, the driver is supposed to release the pressure as a reward for stopping. The problem is most horses that are stopped that way are leaning into the bit when the driver releases the pressure. It's never a good idea for the driver to release the pressure when a horse is pushing on the lines because any release is perceived

by the animal as a reward for doing the right thing.

To achieve your goal of a light, tuned-in, calm look, the driver needs to learn to stop the horse without throwing the means of communication from the lines to an auditory trick. Don't always say "whoa." The driver also needs to learn not to release the pressure on the bit. When bit release becomes the animal's job, the horse will learn to spare himself the pressure.

As the horse travels forward with the driver following behind, there should be no pressure on the lines. To stop the horse, the driver should pick up the slack while walking along, then quit following. A well-trained animal will notice the change when the slack is taken up out of the lines, and he'll immediately stop so that he doesn't cause pressure to the bit. An untrained or poorly trained animal will motor on without regard to this small change in pressure. As the horse continues forward, the lines will tighten and it is the driver's job to act like a post. The horse will likely exert great pressure on the lines as it pushes into the unyielding hands of the driver. As the horse stops its feet, instead of releasing the pressure he has put on the lines by his pushy way of stopping, the driver should maintain the pressure and wait.

Often the driver must learn to be far less reactive. If you lock your elbows at your sides and try to stand heavily without being moved, the horse will soon lose interest in his pushy ways. The driver wins when the horse processes this new way of doing things and decides to fix the problem for himself by backing off the bit. The horse backs off of the pressure by flexing his jaw toward his chest, by backing up his feet, or by doing both. After a few repetitions, he'll make a change in his behavior so he doesn't run into that pressure each time he stops.

People sometimes comment that such silent stopping where the driver holds the pressure until the horse releases his pull on the lines changes the way the horse stops, making it much more responsive in a short time. I point out that the way he stops hasn't changed so much as how he moves. Moving forward while maintaining lightness on the bit is what is being learned. The horse learns to move forward while maintaining respect for the lines. Ultimately, this should be seen as a way of moving forward that maintains respect for the teamster.

DRIVING WHILE HITCHED

When your single or team is hitched, your starting point will likely be from a stop. When a horse is stopped, you should train him to stay still on slack lines. If a driver holds pressure to keep the horse standing, he or she is numbing bit response. Standing still, walking, and trotting should all be accomplished without adding pressure on the lines to maintain the gait. If the horse is at a halt and decides for some reason to leave before the driver gives a definite signal to leave (while the lines are slack), the slack is taken from the lines until he returns to a stop and backs off the pressure. Many drivers apply tension to both lines as a signal to get ready to go forward, which also encourages the horses to tune out line pressure. Well-trained horses at halt will get ready to back up if both lines are drawn back together.

THE START

Instead of yelling their names, kissing, or clucking at horses that are at halt, it is good to start off by picking up slack in a particular manner. To go forward, lightly pick up the slack in one line to show the horse or horses which direction you'll be heading. Even in situations where it is the obvious thing to simply go forward, you should start by pulling only one line. This has the benefit of making the change from halt to start in a distinct manner that only the driver can signal. If your horses go forward when you yell something or when you cluck at them, any person riding on your vehicle or standing nearby can get the same forward impulse that the driver can get, whether the driver wants it or not.

Even if you are going straight down the alleyway of the barn with a team on a wagon, you should pick up the slack enough in just one line to lightly turn the horses' heads. The turn will be so slight when you want to go straight that a casual observer will likely not notice that you even turned the horses' heads. To the horses, it serves notice that a change has been made, and being on only one line, they take that as the forward direction to proceed. Again, the objective is to heighten the communication via the lines and bits. If neither horse notices the change from "halt" to "go forward in this direction," you can begin to lightly kiss or cluck to them until one of them starts moving forward.

The kissing or clucking noise should be slow and steady, each note of sound spaced about one second apart. When one starts to move, you should quit your kissing noise. When the first horse starts to move, it only agitates him if you continue to make the sound. His tightened traces also tighten the traces on the horse that is still standing there, letting the slow one know that he'd better start moving.

GOING WITH REAL LIGHTNESS

To accomplish the goal of going forward with lightness, the driver needs to be willing to heavily restrain the horses' forward movement as necessary. People who think you can accomplish change without pressuring the lines are likely to make their horses much more pushy. Many textbooks on riding horses stress the importance of having "light, following hands" to the point that many a poor student has a hard time getting their equine to ever be truly light on the lines. If the horse learns that your habit of having light hands means you'll never pull hard, even when you should, he'll likely get to be the heaviest pushing animal around. To have a horse that is light on the lines requires having a rider or driver who starts things lightly, then increases the pressure with whatever is necessary to affect the desired change. Willingness to resist the forward-leaning tendencies of a pushy horse is central to making him see the benefit of being light.

Horses that are jumping into flight mode need to be "caught" with line pressure as soon as their heads go up, which is as soon as their mood begins to change from calmness to flight. A split-second of being too slack when the lines should have been tight can mean the difference between a non-event and a catastrophe.

The Most Difficult Change

With this new, relational way of handling lines, the hardest thing for the driver to get used to is the concept of eliminating the pressure when the horse is doing the right thing. He has to show the horse that it can go forward without the lines being tight. For the animal to ever learn the concept, the lines can't be a little lighter than they were. They have to be loose. Most experienced drivers who have been waterskiing behind their horses have to make a conscious effort to serve up some slack. Many teamsters can't make themselves try it. The idea is too strange for them.

The Half-Halt

With most teams and singles, when you first feed them some slack, they take it as a sure sign that they are supposed to take off. It would be a mistake to ever feed slack as a team is pressuring the lines. Usually you can only offer slack as they take their first steps from a halt. When they begin to speed up, immediately administer pressure to the lines and become a post until both horses stop. You should continue to hold the tension it took to stop until both animals yield their heads or back their feet up enough to release the pressure completely. As soon as they process that for a little while, you can start them off again for another go at it. Each time they speed up, stop them without saying "whoa."

A lot of English riders and pleasure-driving instructors expound at length about the advantages of using a half-halt to periodically slacken a horse's stride. If a quiet, gentle uptake of the lines doesn't cause a noticeable slowing of stride, the rein pressure should be increased until the horse stops and releases the pressure. When your single or team noticeably changes to a shorter stride on light line pressure, you can allow them to proceed without coming to a stop. You've just executed a meaningful half-halt. If you pull back on the lines, don't see any change, and then release the pressure for some reason, you've numbed your horse to the cue for a half-halt and a halt.

Going up hills, going around turns, and being hooked to things that are noisy will provide many occasions for the driver to learn to not numb out line response, and it will provide more opportunities for the horses to learn not to be pushy on the lines. For this relational line-handling training, I like to be hitched to something that pulls moderately hard and makes a fairly loud noise. A medium-weight training sled on dirt with one or two people standing on it is usually ideal for a team of mature draft horses. Circular pathways around an arena will warrant a better response in shorter time than going straight forward down the road.

Pushy horses will usually not show much improvement during the first few minutes of being driven in this new way, but they will begin to tune in as the training session progresses. It is important to continue until the horses learn that they can go forward with a load without having to be restrained by the bits. There only needs to be a slight improvement in the horses' attitudes at first, but it's important to see some change to responsiveness before calling it a day. Do not attempt to set a goal in advance for the amount of time you will work on training responsiveness. Instead of thinking, "I'm going to work them for a half-hour," it is better to think, "I'm going to work them until I witness a small change toward lightness."

If you're doing it right, most horses will pick up this light line-handling program in a very short time. They like to be driven in this way because it is relational. Horses are highly relational animals, and I expect they hope you will change from waterskiing on the lines to being relational with them. This way of driving invites horses to use their brains instead of adrenaline to survive. Horses that are responsive have a certain look about them, and they are great to own, great to drive, and great to work with. Whether it is for showing, working, or pleasure, horses that are light on the lines are what everyone really wants.

TURNING YOUR HORSES

When hitched to a vehicle or to a heavy load, a horse moves his body much differently than he does when being ridden or when turned loose. To go around a tight corner in shafts, a horse needs to learn to keep his body straight. If the horse bends his body to get through the turn, it will be prevented by the rigidity of the shafts. The only way for a horse to get around a tight corner in shafts is for him to keep his body straight and scissor his feet laterally as he comes around the turn. Going around the corner with his head and neck out in front is the most comfortable way for the horse to get his feet scissoring as needed. If the head is bent too far around the corner, the opposite hip will jam into the shaft and the horse's feet will stagger around the turn instead of stepping.

The easiest way to keep a horse centered in the shafts and keep his body straight is to handle the reins so the horse can keep his body straight. That means instead of pulling only on the inside rein of the turn, the driver also needs to know how to pull on the

◄ **Head straight and feet crossing.** It's important to know how to turn your driven horses so they can keep their heads out in front of their bodies. Working the feet like this while the head is pulled to the side is difficult for the horse.

▼ **Team turning.** The same head-in-front way of turning is done when driving a team.

▶ **Head is too far over.**
Don't make your horse's head go too far over. If you make your turns in vehicles by only pulling one rein, your horse will look awkward going around turns.

opposite or opposing rein, as needed, to resist head turning. It's a difficult balance for a beginner driver to get the feel: asking for a turn, but then refusing to allow the horse to bend his head and neck in that direction. In many turns, the driver only initiates the turn with the inside rein, then uses the opposing rein to keep the head and neck out in front as the horse goes through the turn.

Gee and Haw Commands

Many draft-horse teamsters place a heavy reliance on the use of the commands *Gee* and *Haw*. *Gee* means "go right" and *Haw* means "go left." I use these commands quite a bit, too, but only when making sharp turns. Some teamsters yell "gee" so their horses will round the end of the arena, but that's a great way to numb the responsiveness to the word. I like to reserve the use of these two words for times when I want the horses to turn so sharp that turning is all they are doing. Some people call it "swinging" their horses when they ask for a turn that comes to the side

without going forward. Used for that kind of turn, it forewarns the horses that this direction you are asking for is not forward, only to the side.

The sequence of pressure that would lead to a sidestepping turn to the left would look like this: While the horses are at halt (they'd be on slack, too), the left line is picked up and the slack is taken out of it. The horses are starting to think about stepping forward and to the left, but then you say "Haw" in an audible, but not harsh tone of voice. You then continue with whatever pressure is needed to swing the horses to the left with your left line as you prevent forward movement with pressuring of the right line, as needed.

COMING UP ON SCARY THINGS

When you are out on the road with a horse, it is very important the horse learns he has a lane where you want him to stay. Obviously, it is dangerous if your horse thinks he can swerve from one side of the road to another to avoid things along the edge of the road

◀ **A sharp turn is dangerous.** This spring wagon was turned as sharply as possible. It doesn't take a very sharp turn set-up for a dangerous situation with this type of vehicle. Notice how a little pressure on the shafts causes the front wheel to go under the body, which lifts the rear wheel as the vehicle begins to tip over.

▼ **Driving on the road.** It's pretty important to teach your horses that their lane is on the right side of the road and near the shoulder. Leaving that lane is not allowed, because shying at a paper cup by the side of the road might get you run over.

that concern him. Drivers of motorized vehicles coming up behind your horse-drawn vehicle usually have the expectation that you will be staying over to the side of the road.

To teach this to a horse, the first thing that needs to happen is that the horse's driver must take all swerving very seriously. Allowing your horse to leave his lane some of the time when you don't hear anything coming up behind you only conditions the horse that leaving his lane will be okay. Horse drivers who have trouble making their equine stay over to the side need to learn to get a bit more intense on the reins as soon as the animal tries to swerve. Some horsemen get the picture when you tell them that their life depends on making the horse stay in his lane.

If your horse seems likely to stop when he can't swerve into the car lanes, then you might want to bring a whip along to deal with that situation. Always remember to kiss or cluck to your horse to encourage forward movement before applying the whip. Horses

▲ **Backing up.** Pressure the reins as needed to get a horse to move backward. It's important to let off of the pressure as the horse backs up. Many people who are driving (and riding) take the non-relational approach of dragging back on the reins without relief until they've gone the distance. That way of backing causes a lot of trouble. Your horse's feet get stuck, his head goes up, the mouth comes open, and rein pressure to get the horse to back up is increased.

seem to handle the whip pressure much better if you give them a fair warning that they'd better get moving first. When a horse finds out that you're serious about going forward, he will usually quit arguing when he hears you make that clucking noise.

Many horse-drawn vehicles that people use on the road have four wheels. In a vehicle like this, if the horse tries to turn around while backing up, the vehicle can be flipped over or the shafts can be broken. It's a good idea to keep your inexperienced horses on two-wheeled carts and four wheeled vehicles that have a fifth wheel for the first few months of road use. These types of vehicles were designed with safety in mind.

BACKING UP

When backing up a well-trained horse in a vehicle, it is important for the horse's driver to know that he or she needs to stay relational. Most people think that the way to back up a horse is to start pulling backward on the reins, then hold them under pressure until the horse has gone the desired distance backward. A better way to back horses up is to pressure the horse as needed to get the feet moving at the desired speed, then release the pressure as the horse's feet begin moving. When the horse's feet come to a halt again, the reins are pressured until the horse begins to come to the rear again. This sort of backing up keeps horses light on the reins because there is a release of pressure when the horse is actually backing up.

Many horse drivers think in terms of a certain distance to the rear as their goal when they begin to seek a back-up on their animal. If your goal is a certain distance to the rear, the horse will be destined to be dragged by the driver for every step to the rear. When the horse learns he will be released on the reins when his feet are moving, he'll start moving his feet at the slightest request for a back-up. Horses that are released while their feet are moving to the rear are learning that it's not about a distance—it's about being respectful and responsive.

Chapter Seven

Longitudinal Alignment

When people run together on a training run, they are supposed to be longitudinally aligned. If you speed up or slow down, it is by agreement, and you do it together. A friend who invited himself along on my daily run wasn't really running faster than I was, he just had to be out in front by about 2 feet for the entire run. If I slowed down, he slowed down. If I sped up, then he sped up. It was maddening. It probably made him feel like he was "beating" me to be out ahead like that. I labeled it "rude."

Not only are people rude to other people when they try to get ahead inappropriately, but horses can be rude to people and horses can be rude to other horses in the same way. These areas of rude behavior differ from human-to-human rudeness because a human can and should affect a remission of rudeness when it is seen in the animals under his care. Not all people believe that people should control horses, but as a horse trainer, I obviously do! Good horse handling makes life much easier and more pleasant for both the horses and the people who are around them. This is a good place to get even in a world where inappropriate forms of getting ahead happen to all of us all of the time.

I've seen many teams of horses ("team" being a clear misuse of the word) in which one horse is being rude. Often in this awkward, inappropriate behavioral situation, the teamster doesn't notice that there is a problem. One horse is rudely pulling on the bit, going ahead of the other, and the teamster isn't doing anything about it. If you are only loose herding a couple of horses out in front of you, one being ahead of the other doesn't matter. Hitch them together as a working team and there is a radical difference in the importance of them working smoothly together. If you want to call yourself a teamster or a whip, you have to learn the skills that will make those two horses become a team in the true sense of the word.

In the chapter about lateral alignment, we looked at the dimension of team alignment that has to do with the regulation of the distance between horses and factors that make them parallel to the tongue at that distance apart. This time, let's look at an equally important part of alignment called longitudinal alignment, which is the way that a pair of horses line up nose to nose as they are working. Ideally, longitudinally aligned horses should be side by side, with an equal feel on the lines, and working so that the doubletrees and neck yoke are perpendicular to the tongue. These two conditions are the main characteristics of longitudinally aligned teams.

LONGITUDINAL ALIGNMENT AND SAFETY

Most teamsters don't realize it, but teams that are not being driven in good longitudinal alignment exhibit hazardous behavior. If a team is walking along out of alignment, it means that one of the horses is not being driven. The behind horse can swerve right or left with quite a bit of freedom. Some horses scratch their itchy bodies by taking a dive at the other horse as they are walking along. Some of the most dangerous out-of-alignment behavior happens when a team is standing still. If one horse is to the rear, that horse isn't in contact with the driver at all. It's easy for a horse that is standing to the rear to reach over and rub his bridle off or tangle up his lines on the harness of the other horse.

While lateral alignment problems are mostly corrected by equipment adjustment, longitudinal alignment corrections are made with increased awareness and through skill of the teamster with the lines and the whip. In more severe cases, one horse may have been allowed to be rude to the other for longer periods of time and he may no longer listen to the bit. In these situations, it becomes necessary to correct the problem with various mechanically applied devices in addition to driving skill.

I have yet to find the team of horses that have perfect longitudinal alignment all of the time without some attention from the teamster. It's the teamster's lack of skill that makes a natural condition (one inadvertently steps ahead of the other) into a problem (one horse has to be ahead of the other all of the time). Given some time out in front, due to inattentive driving, a horse gets to the point that he desperately needs to be in front and becomes rude about it.

FIXING IT WITH DRIVER SKILL

As a teacher of beginning drivers, it is apparent to me that most people don't have one bit of trouble guiding a team through gateways and around prescribed circles. This skill is immediately acquired by virtually everyone. The challenge that takes all of the beginner's attention and quite a bit of practice to attain is longitudinal alignment of the team.

Like a broken record, we proceed around our driving course with me saying over and over to the beginner teamster, "Take hold of both lines evenly. Pull back on both lines until the faster horse comes even with the slower, then ease your pressure." At first, the beginner has the idea that the right line goes only to the right horse and the left line goes only to the left horse, which is contrary to what they see and hear from me. So, as we begin to drift right or left on a straight stretch, they're trying to correct longitudinal alignment by pulling on the line that is on the same side as the faster horse and I go back into my broken record routine.

The lines are designed to slow down the faster horse by pulling back on both lines. When one horse is ahead of the other by an inch or a yard, the one that is out ahead is the only one that feels the tension in the lines. The further ahead the faster horse is, the more you can see the slack dangling on the slower horse's lines. When the team is in longitudinal alignment, both horses equally share the pressure exerted by the teamster. When there are longitudinal alignment problems, the pressure felt by the horse out ahead is exactly the pressure being exerted by the teamster.

Perfect longitudinal alignment with a nice pair of horses (both responding correctly to the same bit pressure) is a skill that all good teamsters possess. If you have one horse out ahead of the other all of the time or most of the time, you either have a seriously rude horse or your skills need improvement. If you suspect it is the latter, then pay closer attention. You have to restrain the faster horse just as soon as you see or feel that he is trying to get ahead. If your faster horse is getting 2 feet ahead of the slower horse before you restrain him enough to slow him down, then your team will be out of longitudinal alignment most of the time. It will also become obvious to the rude horse that he is being successful. If you restrain him and bring him back to longitudinal alignment when he gets 1 inch ahead of the other horse, not only does your alignment look much better, but you don't have so far to bring him back to where he should be.

I have achieved the best results correcting a rude horse by noticing the unwanted advance early and pulling back on the lines firmly, but steadily, so the horse is brought back to alignment over two or three horse lengths. I've seen teamsters who try the "jerk" method of dealing with the rude horse. When they see him out there a foot or so ahead of the other horse, they'll jerk back on the lines really hard and try to make him sorry that he is so pushy. It's logical thinking, but it usually doesn't work. The rude horse often has a more nervous temperament anyway, and jerking on his mouth will only make him more anxious. An anxious, nervous horse will respond to being jerked by getting his head and blood pressure up and he will try even harder to get away from this bad situation.

How Bit Selection Helps

If you suspect that the horse in the lead just doesn't feel you, add more bit. If your horses are wearing adjustable leverage-type bits, then all you have to do is move the lines to a lower position on the shanks of the bit to get more leverage and control. When driving a pair of horses, you can't have one horse pulling 50 pounds of pressure to keep him even, while the other horse won't come up to the bit at all. The

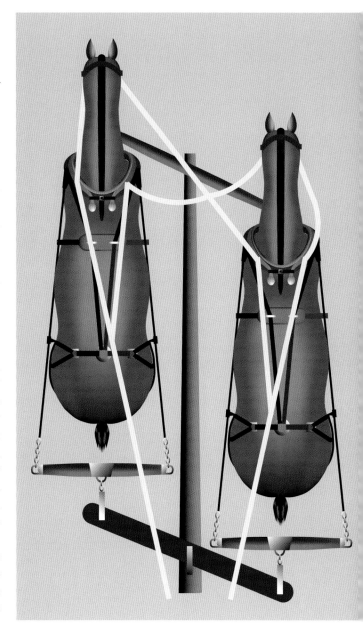

▲ **A longitudinal alignment problem.**

amount of pull on the lines on a pair of horses should be about the same for each horse or you will definitely have longitudinal alignment problems.

The amount that a horse desires to go forward can change even during a single work session. When big (multiple-horse) show hitches have been put through a long trotting session by the judge and they come into the center for a time, the attendants come running out to put the lines up on the bits of some of the horses (they've become tired and less energetic) and maybe down on one or two of them (the crowd or something else is exciting them and they are leaning too hard on the bit). When you are driving more than two horses, it becomes obvious that you must have all of the horses fairly equal in how they pull on the bit.

The ability to quickly change from mild to more severe, or vice versa, is the reason Liverpool-type leverage bits are popular for driving. As a word of caution, you need to be aware that the horse will tell you how much bit he needs. You can't just arbitrarily decide that all of your horses are going to be driven in this bit position because it makes you happy. You aren't the one wearing the bit, the horse is! If your bit selection is too mild, some horses will pull you right out of your seat by their mouth. If your bit selection is too severe for the horse, you might get to see what "balky" looks like. Generally, as bit severity increases, pulling confidence (on the lines and the load) decreases. There is a need for experience and sensitivity on the part of the teamster to properly respond to information the horse is sending about bit requirements.

Should I Speed Up the One or Hold the Other?

Encouraging longitudinal alignment by using the whip to move the slower horse forward is sometimes needed. Being behind all of the time and not being in contact with the teamster destroys the initiative of a horse. Some need a little more ambition, which is provided by the whip. In team driving where longitudinal alignment is important, the slower horse is always the one that is setting the pace. If your ground speed is too slow when you have them aligned longitudinally, then you know that the slower horse needs to speed up. Making the slower horse's bit more mild is the first thing to try (wrap the mouthpiece of the bit with rubber, if necessary) if the faster horse

is responding well to the bit. This is often all that is required to get a team into longitudinal alignment.

I've heard many teamsters say that they'd use their whip a little more on the slower horse if it didn't make their faster horse go even faster. The sound of the whip striking the slower horse is enough to speed up the fast horse even more. Maybe a goad is a solution in that case. I had a pair of mares who had that problem. A helpful rancher with his pitchfork showed me how to quietly and quickly fix the one's motivation problem.

Can't figure out which horse in your team is being rude? If you're thinking that maybe it is the slower horse because if he'd just get going as fast as the other horse, they'd be in longitudinal alignment, get your whip out and give it a try. If as you speed up the slower horse, the faster horse goes faster, then you have a clue. If you have trouble getting the faster horse to yield to the lines, you have another clue about which horse is being rude. It's been my experience that 95 percent of the longitudinal alignment problems are caused by the faster horse being unwilling to yield to the lines. If the faster horse is pulling 80 or 100 pounds on the lines, there is no way that the slower horse will be able to relate to that kind of pressure. As he tries to come even with the faster horse, he moves from a position that has zero pressure on his bit to one where there is extreme pressure. It would be like trying to push against an iron railing with the soft parts of your mouth. The slower horse will invariably keep himself where he doesn't have to come up against that pressure. My sympathies are usually with the slower horse. Even if he really wants to get even with the other horse, he can't because it hurts too much.

Is it Because of Unmatched Stride?

Often I run into people with longitudinal alignment problems who dismiss the problem by saying, "The reason I have this problem is that one of my horses has a long stride and the other doesn't." To that statement I always reply, "It doesn't matter about their stride." I want a horse to keep himself where he belongs (even with the other horse) no matter what. At the same time, I will try to do things to keep their difference of stride less noticeable. If a fast walk by the longer-strided horse causes the short-strided

▲ **Using Liverpool bit line positions.** The farther down the bit the lines are used, the more severe the leverage. The top position, the snaffle position, is where the line is hooked into the big ring level with the mouthpiece. The next one down is called the "lady's curb," which is directly below the mouthpiece. The next after that is often referred to as the middle curb position. The bottom slot on the cheek piece, called the bottom or full-curb position, is the most severe.

horse to trot or jig, I will slow them down enough to keep the short-strided horse walking. First of all, one horse walking and the other trotting doesn't look good. More importantly, when a horse is jigging he is not really using himself well. Smooth, calm movement with a free-swinging back and each foot placed confidently and firmly on the ground is the way a horse's gaits (at walk and trot) should look. When a horse is allowed (or trained) to mince along with the nervous, short, stiff-backed, low-strided trot

called jigging (some people call it their parade gait), a mistake has been made. Horses that jig normally are not good pullers. I haven't quite got it figured out yet, but either jigging makes a horse nervous or being nervous makes a horse jig. It's kind of like the old chicken or the egg question. Which came first? Part of your job as a teamster is to prevent development of unwanted gaits. If you have a horse that likes to jig, you have a problem. Sometimes jigging horses will quit this behavior if their bit is softened.

▶ **Short neck/ long neck adjustments.**
To hitch a short-necked horse to a long-necked horse, make the following adjustments: Both the draft and the coupling line move forward for a long-necked horse, and the lines are brought an equal distance to the rear for a shorter-necked horse. This takes care of the neck length problem without affecting lateral alignment. Remember that neck length issues can be affected by how the horses carry themselves when working.

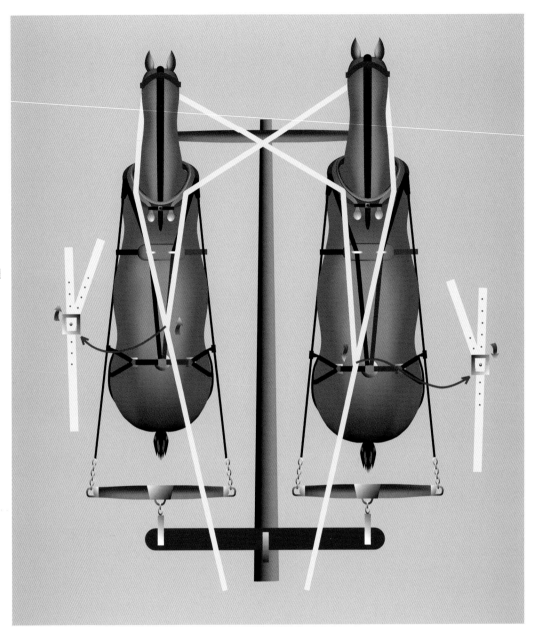

Short Neck/Long Neck Adjustments

Occasionally you will run into the problem of having to hitch a short-necked horse with a long-necked horse. Also, some horses get ahead by flexing or overarching their necks enough to get the doubletrees ahead on their side, even though both horses' necks are actually the same length. Whenever your driving or adjustments cause the doubletrees and neck yoke to not be perpendicular to the tongue, you have violated one of the two basic rules of proper longitudinal alignment (equal feel on the bit, and doubletrees and neck yoke perpendicular to the tongue). That's why it's not good to monkey around with differentially adjusting the trace length. In either case (different length necks or an overarched neck), you assume that the horses are an equal distance ahead of the doubletrees with their collars (the horses have to be wearing harness with equal length traces to make that assumption).

This way, even if you are hitching a draft horse next to a pony, they will still have their doubletrees and neck yokes longitudinally aligned (perpendicular to the tongue). To make sure the shorter- and longer-necked horses have an equal feel on the bits when the doubletrees are perpendicular to the tongue, you will have to make some adjustment to the coupling line buckles. This adjustment will accommodate the difference in their length from the collar forward to the bit, and the adjustment must be symmetrical to prevent either horse's bit from being tilted.

If there is a 6-inch difference in neck length (actual or due to over-flexion) and the draft line is punched every inch for adjustment of the coupling line buckle, then the coupling line going to the longer-necked horse is moved forward six holes. The draft line going to the longer-necked horse is lengthened by six holes (the coupling line going to the shorter-necked horse is moved back six holes). With this adjustment, you are simply allowing for a longer neck on one horse and not affecting lateral alignment. It is a symmetrical adjustment that doesn't tilt the bits crossways. This is a useful adjustment to assure longitudinal alignment in otherwise difficult circumstances.

The Wooden Doubletree Effect

Why is it that some horses want so desperately to be in front? The thrill of victory motivates horses, but there is another factor on wooden doubletrees that causes horses to get ahead. When using wooden doubletrees, because of the way the center draft hole and the holes for the attachment of the singletrees are offset to prevent breaking the wood, there is some advantage to being out in front. As the doubletree swings toward the horse that is in the lead, the distance from the hitch hole to the singletree hole lengthens on that side and it is shortened by the same amount on the slower horse's side. Is there enough of a change that the horse in front can feel the load get lighter as he gets out of alignment? Horses seem to be very sensitive to these things. Steel doubletrees usually don't have an offset to the holes, so swinging the doubletrees shortens the length from draft hole to singletree hole equally on both sides.

If you think it would be difficult to get ahead of a rude person on a training run, try getting ahead while

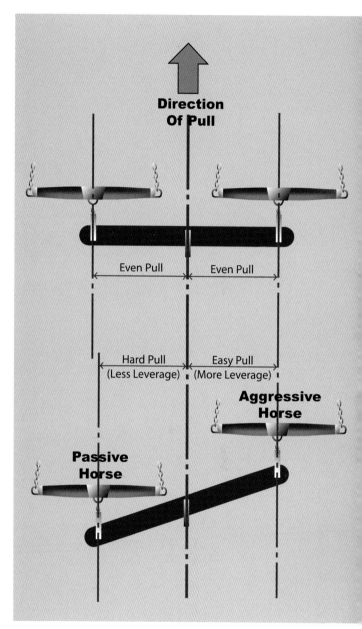

▲ **Wooden doubletree effects.** The dotted lines mark the center of the holes on a wooden doubletree, making it easy to compare their distance apart when straight and tilted. When the doubletree is tilted due to one horse trying to get ahead, the distance between the holes doesn't stay equal. Longer distance to the center hole on the aggressive horse's side makes his load lighter.

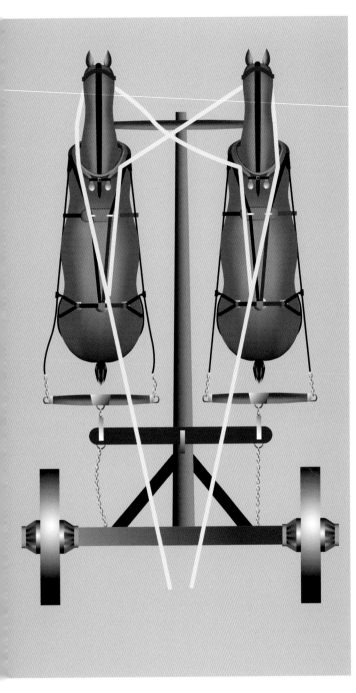

▲ **Stay chains.**

you have a doubletree load to share. You not only have the ground to make up (the distance between the two of you), but you also have to pull more than half of the load (if you are on wooden doubletrees) as you gain the ground. I think that is the reason poorly driven teams will usually be made up of a pair of characters you could name "Aggressive" and "Passive." If "Aggressive" isn't restrained, "Passive" will never get even. Their characters become very well set.

You don't need to get rid of all your wooden doubletrees to quit making it easier on your faster horse. A better solution would be to drive your horses well aligned on both kinds of doubletrees so that they will never experience the unfairness that being out of longitudinal alignment produces.

MECHANICAL REMEDIES

Some horses have been allowed to be in the lead for so long that they become too much trouble, even for an experienced teamster, to correct without some sort of mechanical device. If you are pulling on the lines, but not getting the response of the faster horse coming back to you (even with a leverage-type bit in one of the more severe positions), then you either need some mechanical help or extensive retraining of bit response. The remainder of this chapter deals with the somewhat rare circumstances where bit and line adjustments and skilled driving are not sufficient for bringing a team into longitudinal alignment.

Stay Chains

The most common mechanical remedy for longitudinal alignment problems is to use stay chains on the doubletrees and make the aggressive horse pull the whole load. This is easily done and it does work; the horse tightening the stay chain can pull the whole load if he wants to, and most of them do want to. I never stay-chain my way to longitudinal alignment because instead of making the problem go away, I feel it makes the problem worse. The rude horse really has a problem with being overly motivated. If you raise the requirement of work, he will respond by becoming even more motivated to pull, making his rudeness even worse. The heart rate, respiration, desire to go forward, and pressure on the lines all increase together.

Stay chains are useful when properly employed as a safety device. As a safety device, they are used to prevent a very unwanted longitudinal alignment problem that I call "shear." In other words, they prevent one horse from bolting ahead of the other. When one horse leaps ahead of the other, he can get so far ahead that he jerks the coupling line buckle on his coupling line through the upper hame ring on the other horse. If this happens, no amount of pulling by the teamster can get it back through the ring, and the teamster has effectively lost the ability to stop or turn either horse in that one direction. Even if that little problem doesn't happen, shear is still bad because it is usually the first step to a runaway situation. Stay chains prevent shear because they do not allow extreme swings of the doubletrees. Adjusted for that purpose, they are a great safety device.

Wear 'Em Out?

Some teamsters deal with longitudinal alignment problems by increasing the load being pulled by the team to try to wear out the aggressive horse and teach him the error of his ways. I think that it is easier to correct these problems on light to moderate loads. With a team that is out of alignment on a wooden doubletree, the passive horse is pulling more of the weight anyway, and then is being asked to step up even with the aggressive horse. It's much more easily done on a lighter load. The aggressive horse won't be pulling on the bit as hard if he is on a lighter load. The heavier the load, the harder a poorly trained horse pulls on the bit. On a light load, you'll have a lot less trouble restraining the aggressive horse.

The Tie-Back

The simplest, quickest, and easiest way to restrain a horse is to simply tie him by a lead rope attached to his halter to some tie place on the harness of his partner. I use this method of assuring longitudinal alignment for cases in which alignment problems intermittently occur, such as when driving a colt with a breaking horse. The further back on the other horse that you make your tie, the more turning ability the restrained horse will have. I usually tie to the bellyband billet/trace junction. While it is very strong and effective at restraint, the drawback to this method is that it doesn't allow sharp turns in the direction that the restrained horse is occupying if that horse is leaning hard and constant on the tie back rope. The restrained horse, if he is on the right, can't turn sharply to the right in that situation. I don't use this method of longitudinal restraint when I am dealing with chronic aggressive behavior.

The Buck-Back

Another restraint system is to "buck back" the aggressive horse to the other horse's trace. As the aggressive horse tries to step ahead, he pulls his side of the doubletree forward with his traces. The passive horse's side swings an equal distance backward. The buck-back system harnesses that motion of the doubletree so that the harder the aggressive horse tries to get ahead, the more he will be pressuring his mouth. When he responds properly to the pressure and moves back into longitudinal alignment, the pressure is appropriately relaxed. This system also allows the restrained horse to turn as directed by the teamster because the free running ring that the buck rein passes through allows the head to turn.

The buck-back rein I use is a 1-inch nylon tape that is about 14 feet long. It has a rein snap on each end. This rein forms a big loop, starting at the bit on one side of the mouth, going through the upper hame ring on that same side and through the upper hame ring on the other side of the horse, and ending at the snaffle ring on the other side of the mouth from where you started. Between the hame rings, a strong harness ring of about 2 inches in diameter is inserted. This ring has a nylon strap about 4 feet long attached to it, with a heavy-duty snap at the other end. The snap is used to fasten the strap to the inside trace of the other horse.

On older, aggressive horses that need to be restrained to achieve longitudinal alignment, I'll fasten the buck-back rein into the snaffle position on a leverage-type bit. I never hook into a more authoritative position than the snaffle with the buck-back system because I want to use the more powerful positions for the driving lines. Thus, I can override the longitudinal restraint to steer and stop.

There are many positive things about this system. If you've been going along pulling so hard on the aggressive horse to keep him back where he belongs that you swear your arms are getting longer, you will now have complete relief. The buck-back system makes longitudinal alignment problems go away. All you have to do is steer. It is driving the way beginners expect it to be.

I put the buck-back system on both horses if there is any danger that they might try to run off. It's "the great equalizer" not only of horses as they work side by side, but also in the sense that it makes it a lot more possible for a teamster to prevent runaways. I've noticed that runaway teams usually start off by first getting severely out of longitudinal alignment. One horse gets spooked and leaps sharply ahead, dragging the poor teamster forward with the lines. The other horse, which may or may not be of good character, responds to the excitement and the completely slack lines by leaping forward as well. As he rockets past the other horse that started it all, he drags the lines forward on his mouth, giving the instigator the slack lines he needs to leap ahead again. They shear past each other like that a few times until there is no difference in the speed of either. They get over their longitudinal alignment problems as they approach runaway speed. That is not a safe way to achieve longitudinal alignment!

▲ **The buck back is a restraint system.** A 14-foot-long rein with a snap on each end is attached to a horse. The rein passes through the upper hame ring and goes to the snaffle position on both sides of the horse. The short piece with the ring that attaches to the other horse's trace is being held in the teamster's hand.

◀ **How to hook to the trace.** To hook to the trace, hook the end of the short piece into the passive horse's inside (medial) trace. This short piece can be lengthened and shortened by changing the toggle chain link it is fastened into. It can also be shortened or lengthened with the buckles holding the ring and snap.

With both horses bucked back to the other, they have a hard time getting anything going. Both horses are forced to keep an even feel on the lines no matter what, as they are supposed to do. A little time spent being bucked back teaches a horse to be careful about leaping ahead.

WHAT CAUSES THIS PROBLEM?

You might ask if these chronic longitudinal alignment problems are always caused simply by the teamster being unaware or uncaring about this dimension of alignment. I think that the problem most often starts that way, but then it quickly progresses to the next inappropriate thing that a beginner will do. They see that they have one horse wanting to be out ahead all of the time and they sense that pulling on the lines seems to help a little, but they just don't know enough to stay relational with their line use.

If a teamster forgets to be on and off of the pressure, depending on how the aggressive horse is acting, the horse will end up being held constantly with every step he takes. This is where the phrase "impossibly hard mouth" comes from.

A beginner must learn to smoothly and quickly increase line pressure in response to hills (or any other load increase), noisy equipment, sudden loud noises,

▶ **In use.** When the aggressive horse tries to get ahead, he drags his singletree forward, which scissors the other singletree to the rear. The buck-back uses that swing of the doubletree to pressure the aggressive horse as he steps ahead of the passive horse.

and other factors. You have to be strong enough to keep the horses walking in those situations. Also, the beginner needs to learn when to immediately ease off the pressure, such as when the faster horse comes back to longitudinal alignment or when they slow down in response to the lines.

Using the buck-back system to obtain longitudinal alignment is not as good as careful training and awareness to prevent such problems from occurring in the first place. Many horses are never taught how to lightly respond to bit pressure when they are first trained. Assuring lifelong proper response to bit pressure (and to other kinds of pressure, such as collar, whip, trace, noise, etc.) requires careful training and careful use. I am a firm believer that training by using, which is common with draft horses, is not really training at all and

is the cause of most behavior problems. Gentle and progressive training of good response to the pressures put on the driven horse (isolating each of the above pressures and teaching proper response to each pressure in a logical and sequential manner) assures right and light response for the rest of the horse's life.

It is sometimes frustrating living in a world where rude behavior is everywhere. Fortunately for those of us who work with animals, here is a place where we can have some influence. Here we can and should assure fairness. The next time you have a "rude person experience," don't get mad. Resolve to uphold fairness and good manners in an area where you have some influence, such as with the horses you drive. It'll make you feel better about yourself and make your horses happier, too.

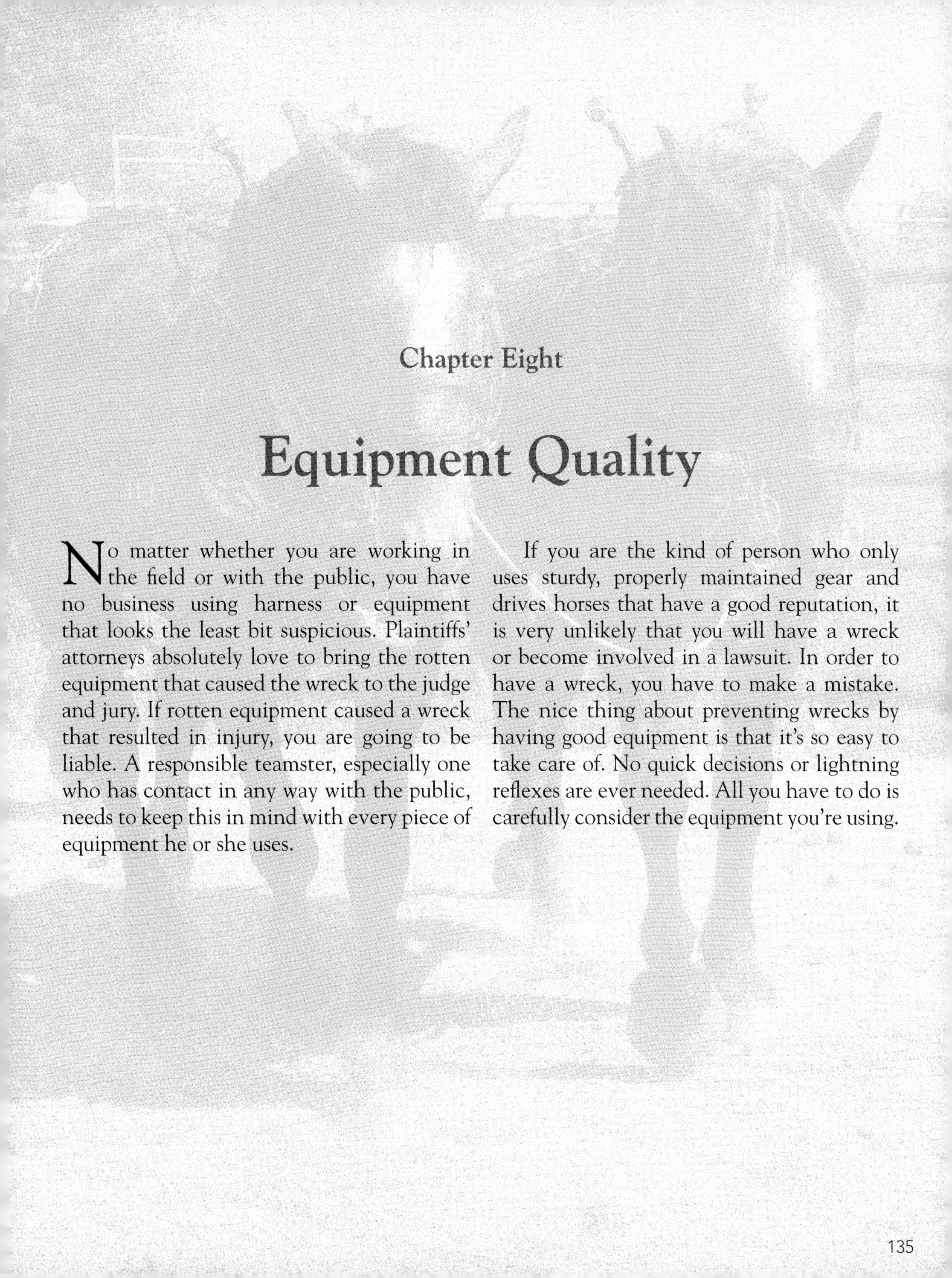

Chapter Eight

Equipment Quality

No matter whether you are working in the field or with the public, you have no business using harness or equipment that looks the least bit suspicious. Plaintiffs' attorneys absolutely love to bring the rotten equipment that caused the wreck to the judge and jury. If rotten equipment caused a wreck that resulted in injury, you are going to be liable. A responsible teamster, especially one who has contact in any way with the public, needs to keep this in mind with every piece of equipment he or she uses.

If you are the kind of person who only uses sturdy, properly maintained gear and drives horses that have a good reputation, it is very unlikely that you will have a wreck or become involved in a lawsuit. In order to have a wreck, you have to make a mistake. The nice thing about preventing wrecks by having good equipment is that it's so easy to take care of. No quick decisions or lightning reflexes are ever needed. All you have to do is carefully consider the equipment you're using.

▲ **Fresh paint.** Most teamsters prefer their wooden hitching equipment to be clear-stained and oiled or varnished instead of painted like this neck yoke. The smoothness of the finish and the fresh paint make me think this one was probably smoothed up with body filler, so the first time you put a little pressure on it, the truth will be revealed.

▲ **A wheel that's ready to go.** The worst thing about using a wheel like this is how it feels when the wheel keels over while you are going around a turn. The one time I saw a wheel break, it was on a cart going around a turn on a hillside. The driver, a big fellow, ended up doing a body slam onto his passenger as they both hit the ground. Ouch!

▲ **Old, dry harness.** It's a lot cheaper in the short term to buy a used (and often old and dry) harness instead of a new one. But this kind of harness is ultimately the really expensive kind, which teaches a lesson that is usually learned after the wreck.

THE POLE STRAP LOOP: A SAFETY BRAKE

Quite a few draft-horse team harness have a loop in the end of the pole strap that is often ignored by teamsters. The function of this loop is to act as a safety brake—a back-up brake system—in the event of failure of other harness parts farther back. By having the bellyband slung through the loop in the end of the pole strap, instead of passing the bellyband under the pole strap, the level of safety is increased greatly. This use of the loop engages the strong holding ability of the backpad, market tugs, and bellyband that encircle the horse (the surcingle) to prevent the load from overtaking the team in the event of any failure in the quarter straps or breeching.

Picture a team of horses holding back a heavy wagon or sleigh on a steep hill without the bellyband through the pole strap loop. If a quarter strap breaks or a ring is ripped out of the breeching, the breeching is pulled away from the side where the break occurred and it is pulled forward around the horse's rump by the unbroken side. This sudden repositioning of the breeching to one side of the horse makes for too much length in the braking system of the harness. At this point, the eveners contact the hind legs of the horses, and the plot thickens. This sort of story doesn't have a happy ending.

However, with the loop in the end of the pole strap engaged, the weight of the load will be smoothly and ably caught by the backpad, market tugs, and bellyband that encircle the horse behind the withers. The horse will no doubt be surprised by the sudden shift of the weight from the breeching to the surcingle, but nowhere near as surprised as he would be at the feel of the load crashing into his hind end.

To illustrate this function of the pole strap loop to yourself, disconnect one quarter strap from a harnessed horse and pull the coupler snap forward, first without the bellyband passed through the loop. Notice how easy it is to pull the coupler snap up near the horse's chin. Leaving the quarter strap down, try this again with the bellyband passed through the loop. If the harness is properly adjusted, the coupler snap will only be able to come forward about 4 inches more than normal.

When we explain this function of the pole strap loop at our driving workshops, someone will usually

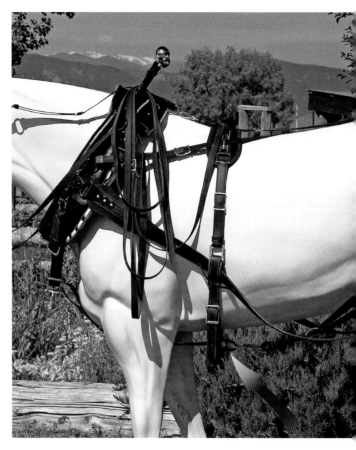

▲ **The pole strap loop.**

say you shouldn't be using a harness that will give way in the breeching or quarter straps in the first place. Of course, that is a true statement, but it is no excuse for not using a safety device that is so easily employed. Spares kits are required on all turnouts at combined driving events. Carrying a complete repair kit does not imply that one is using a rotten harness. The point is that harness breaks do occur for a variety of reasons and that prudent whips and teamsters do all they can to limit the effect of a break. Sure, it takes a little extra effort when harnessing to get the bellyband tucked through there where it belongs, but you'll be happy you did if you ever need the safety brake.

THE KICK STRAP

My friends had a great old horse, retired now, but still as useful as ever. They just didn't use him nearly as much as when their family hauled him all over the country,

▶ **Kick strap.** The kick strap on this harness is attached to the back strap where the strap forks for the crupper. If your harness doesn't have a crupper, the kick strap is somehow passed through the back strap near the rear end of the horse. It's important that the kick strap is attached so that it can't migrate forward or back when in use.

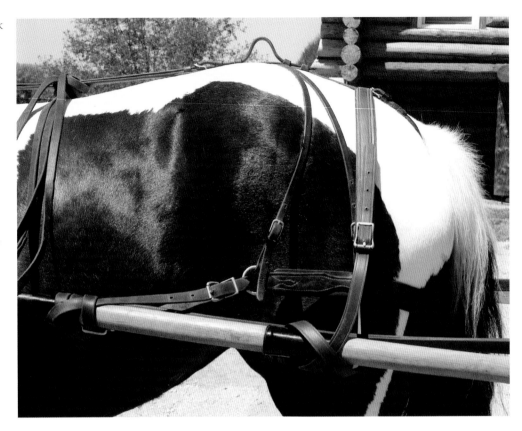

winning halter, riding, and driving championships at every major show of his breed. It was a pleasant summer evening, so they got the old horse out and went for a leisurely drive around the neighborhood to enjoy the peace and security of driving such a great horse.

They had stopped to talk in the driveway of a friend when the unthinkable happened. This perfect horse, which had been standing like a statue, suddenly went berserk. He put his head down and began to violently kick the cart apart with his back feet. One of the blows struck the man in the side of the head, and another got the woman in the arm before they managed to lurch out of the way. Oddly, the horse never took a step forward as he blasted the front out of the cart.

An ambulance was dispatched for the people and their vet was summoned because the horse had cut himself badly. "He was cut, bruised, and bleeding from the hocks down, and I almost didn't see it," the vet said. "I happened to look up toward his flank, and there it was. His sheath was swollen up to about

twice the normal size. The poor guy had been stung by a hornet, right where it really hurts." A kick strap would have made all of the difference and prevented this wreck.

A kick strap is a piece of driving-horse equipment that is considered an accessory, but around our training farm, you'll find that the kick strap is used on every horse in single harness. It's a harness accessory that, when correctly applied, virtually eliminates the chances of getting yourself or the vehicle kicked. Contrary to the indication given by its name, the kick strap is actually an anti-kick and anti-buck strap. It is a beautiful experience to be sitting behind a frisky colt that is trying to buck when he is wearing a kick strap. He might be squealing and bawling like a saddle bronc, but noise is the extent of the damage. He can't do any harm because he can't get his rear end off the ground when the kick strap is in use.

To understand how a kick strap works, one needs to study the body movements of a horse at play. In order for the rear hooves of the horse to reach

◀ **Here's how the kick strap functions.** The pivot point of the lever (the shafts) is at the tug loops. The shaft shackles prevent the pivot point from moving up, while the tug loops prevent it from moving down. The horse is given a short section of the lever bar (the distance from the tug loop to the kick strap) to try to lift the shafts when kicking. The driver, vehicle, and passengers use the lever bar to hold the horse's hind end down.

rearward and upward, the hindquarters of the horse must be elevated. If the upward mobility of the rump is restricted, then the kicking ability is also restricted. The kick strap is a clever device because it uses other parts of the harness, shafts, vehicle, and even the weight of the driver and any passengers to accomplish its effect. When the horse tries to kick up, he becomes the operator of a lever. The horse is disadvantaged because he has been given a short lever in order to lift something that is being held down by a longer lever—the driver and the vehicle.

The shaft wraps (i.e., overgirth, shaft shackles, etc.) need to be stout because they anchor the end of the lever bar (shafts) to the horse when he kicks. With a high-quality lever (shafts), the horse will not normally be strong or heavy enough to lift the load at the other end of the lever (driver and the vehicle), thus preventing the kick.

In my opinion, there are good reasons for using a kick strap. It is effective when properly applied; it is inexpensive (usually less than $80); it is humane to the horse; it inconspicuously blends in with the rest of the harness; it is easy to use; and it greatly reduces the risk of a serious wreck. We often call the kick strap a "cheap insurance policy." Of course, it's not like insurance at all. It's better, because it prevents injury from ever happening.

Many drivers do not have to deal with horses that frequently display difficult behavior, but even the best of horses can get to feeling a little too good. Too much confinement, good feed, and brisk weather can come together to make even a plug feel like playing sometimes. When any horse, whether a beginner or well trained, kicks up and hits the floorboards and singletree area of a vehicle, it probably won't quit kicking until it has removed itself from the vehicle. That is why the kick strap is standard equipment for us. A playful buck or a reaction to being stung is barely noticed if a kick strap is used.

Many drivers rely on a shortened overcheck to accomplish the same safety advantage that a kick strap provides, but a jacked-up head hurts a horse's pulling ability and is inhumane because of the way it causes a horse to move so artificially. Leaving an overcheck fastened while a horse is parked somewhere is cruel and unnecessary.

This simple accessory, the kick strap, can greatly reduce the danger level for both the horse and the driver. It can help change the outlook of the person who says, "I just don't feel totally comfortable being right there where I could get kicked." It's an understandable concern for many drivers. When the danger level goes down, the confidence and enjoyment levels go up, both for the human and the horse.

How to Use a Kick Strap

In the sequence of putting to, the kick strap is attached to the vehicle last, after securing the shafts and breeching wraps on a cart harness. Some vehicles are fitted with a footman loop (metal dee or staple) on the rear area of the shafts where the kick strap goes through, but this loop is optional and we generally do not have these on our vehicles. Do not try to run the kick strap to the shaft loop for the breeching because these are too far forward on the shaft.

Assuming the kick strap is attached to the harness and the horse is completely attached to the vehicle, the steps for attaching the kick strap are as follows:

1) The kick strap should be hanging down by the horse's side between the breeching and the trace. Pick up the end of the kick strap and drop it down over the shaft, wrapping it from the outside to the inside. The shaft only is encircled by the kick strap, not the shaft and the trace.

2) The end of the kick strap is then buckled in the appropriate hole. To find the right hole, gently push the horse's rear end away until it touches the shaft on the other side. Buckle in so that the slack is just taken out of the kick strap. If it is adjusted to be too tight, it could interfere with a horse's natural movement, especially if its use roughly jerks the crupper to the side when turning. If it is too loose, you might as well leave it off.

At times, when we've been in need of a kick strap, we've simply used a lead rope, tying it to the shafts and passing it through a loop in the back strap so that it stays where it belongs. The ready-made kick straps we use are 1 inch wide and 10 feet long, end to end. The kick strap has two layers of leather between the buckles, which are 37 inches apart. At the center of the kick strap, there are slots where the back strap passes through. The slots are formed by leather "shims," or spacers, within the strap. The kick strap we use is made to slip onto any single driving harness at the posterior end of the back strap, where it forks to accommodate buckling the crupper. A kick strap that is made with only one slot is intended for the back strap to fit through, just in front of the fork.

NEARLY A RUNAWAY

Making the final turn on the old horse-drawn hay rake, I was counting myself lucky to have gotten the field raked with this flighty pair of young mares. They were a pair that had been worked before but were then given a couple of years off, and I was just starting their new training. Their owner wanted me to give them a real job again after their long vacation. In the next instant, due to using normal length stub lines, they almost became permanently retired.

One of the mares was spooked and shot forward. Her rocketing forward jerked the other mare backward an equal amount, as the evener bar on the doubletrees pivoted about its middle. The stub line buckle fluidly shot through the upper hame ring of the other mare. When I tried to pull the stub line buckle back to the rear side of the upper hame ring, it hung up on the ring and wouldn't come through, effectively fouling the further use of that entire line on that side. These mares wanted to go toward home fast.

Acting instinctively, I pulled the still-functional other line to put them into a tight circle, as I baled off the rake into the center of the circle. I jumped in front of them, raised my hands into their faces and shouted, "Whoa!" as they were about to break into a run. They almost didn't stop! Standing there, suddenly out of breath, looking at those two spooked horses with their heads up and that big steel rake reel still spinning behind them, I had a brand new set of questions about stub line length.

STUB LINE LENGTHS

In European countries where coupling reins are used on teams and pairs (their farming teams are also often worked in a loop rein, which is an entirely different

To adjust the length of the kick strap, push your horse's rear over against the far shaft. As the rump of the horse touches the far shaft, adjust the kick strap so the slack is almost all removed. This adjustment allows the kick strap to be fully functional, but it's not so tight that it irritates the horse.

reining system), coupling lines are significantly longer than those commonly used here in the United States. Their lines are long enough that the driver can simply reach forward from his or her position on a carriage or wagon and adjust the lines with ease. Easy adjustment had been my reason for the longer coupling reins until that incident with the hay rake. Standing there in front of the rake, it occurred to me that if I had been using European-style coupling reins with that extra length, my near-wreck wouldn't have happened.

The lines I was using had stub lines that were about 5½ feet long, a common length for American-made stub lines. European drivers typically use coupling lines that are about twice that long. Since our driving culture here is descended largely from European ancestors, I began to wonder why the typical lines made in the United States are so much shorter. I began to see this as "The Case of the Disappearing Stub Line."

Why So Short?

What sort of plot was this that was so endangering American teamsters, and who was behind it all? Were short stub lines on this side of the ocean more desirable for some reason? As a teacher of driving skills, I have considered these questions. It seems to me that short stub lines answer to only one motivation: the idea that using less leather is cheaper. To save a dollar or two in leather (or nylon or beta) strapping, teamsters have been duped into driving in lines that are literally an accident waiting to happen. Harness makers do well to save money wherever they can. That is how we cut

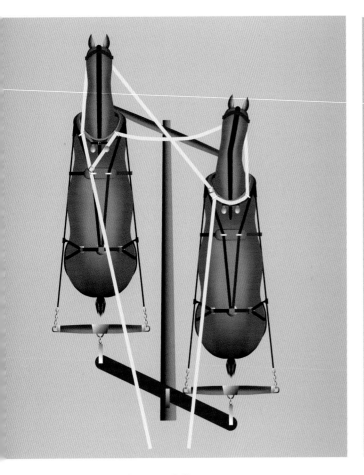

▲ **Shear with short stub lines.**

▲ **Short stub lines.**

costs and increase profit, one of the basic principles of making money. But if harness makers and their customers realized the real cost of cutting material used in the coupling lines, they'd be happy to try to save money some other way.

If you look at the path that a set of team lines take from the teamster's hands to the bits on the horses, you'll see that short coupling lines are not an asset, but a liability, even for normal stopping and steering. The shorter you make a coupling line, the more it tends to deflect the draft line when the lines are pulled. Deflection of the draft line causes warped characteristics in the driving lines that are not desirable. If your coupling line is so short that its use deflects the draft line, you are adding an unintended variable into the relationship between your hands and the horses' mouths.

Living with Short Stub Lines

In the United States, the populace has come up with a variety of schemes to dampen the dangers of living in the Land of the Disappearing Stub Line. The use of stay chains on the doubletrees prevents the extreme swings of the doubletrees that pull stub line buckles through hame rings. Stay chains are a pretty good safety measure, but they cost so much to install on all of your vehicles that I find them impractical and unnecessary. If you are ground driving or skidding logs, you won't be able to use stay chains anyway.

Another approach to preventing the wrecks that usually follow short stub line hang-up is the use of roller-type spreaders. Roller-type spreaders are tight enough to not allow the stub line buckle through. This idea and the buckling in of large rings at the draft line/stub line junction (rings that are larger than the upper hame ring, thus keeping the buckle out of the hame ring) are seemingly good ideas, too. If one of your horses is rocketing forward and jerking the other horse backward, are your lines going to be strong enough to sustain the sudden impact of about 4,000 pounds hitting that spreader roller or hitting your safety ring? Even if everything holds, the wrenching impact of being stopped by such devices is not a nice thing to do to your horse's mouth.

I've been told that the primary reason some teamsters who only ever drive on road-length and shorter doubletrees and neck yokes (48 inches or less) are using line spreaders is to try to alleviate the line hang-up problem. They are using line spreaders because the ring at the end of the spreader is often much larger than the upper hame ring, and large enough that the stub line buckle can go through and back without hanging up. When they are inappropriately using spreaders, they need to adjust the stub line buckle a considerable distance toward the driver's hands to get the bit centers lined up with the doubletrees and neck yokes, which helps keep the stub line buckle from hanging up.

Unfortunately for the horses, most teamsters who use spreaders (when they don't really need them) are not aware that the bit centers need to be adjusted to match the doubletrees and neck yoke. Most teamsters who are inappropriately using spreaders aren't bothering to adjust their lines so that their horses' heads are lined up with the direction of travel. It's not uncommon to see teams in fancy line spreaders with their heads flared out to the sides as they try to go forward. Such poor alignment causes much upset and discomfort for the horses. Even if the stub line buckles are adjusted to compensate for the use of spreaders, you are using an accessory apparatus that provides plenty of extra "looseness" and potential for failure in most cases. There's no need to add on those extra pieces of equipment when using the proper length stub lines will solve the problem.

Adding Length to Your Stub Lines

The lines I use and sell are made with stub lines that are about 8 feet long. When I receive a used set of lines that have a bad case of the disappearing stub

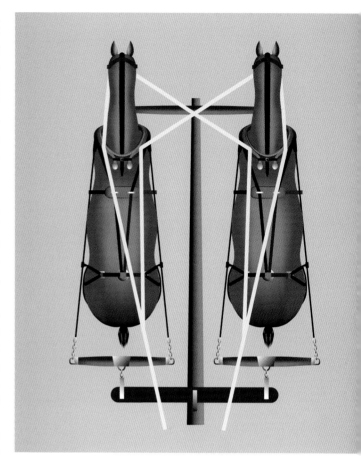

▲ **Long stub lines.**

lines, I'll have my harness maker add 2 or 3 feet to the existing stub line. You'll have to re-punch your draft lines to account for the increased length of the stub lines, but that is simple to do. From the stub line buckle, hold the draft line and the stub line together as you slide your hands forward to the bit end of the lines. Arriving there, I want to see the stub line being equal in length to the draft line. I go back and forth between the stub line buckle and the bit end of the lines until they are of equal length. I then mark the point where the buckle is on the draft line for punching. Ahead of that mark, I make 16 more holes that are spaced 1 inch apart. This allows for adjusting the bit centers anywhere from 26 to 58 inches apart. Theoretically, the actual effect of line adjustments will vary among teams. The variation is caused by differing neck thickness, with the center hole at

▲ **The breeching tie.** The breeching tie should only be functional while driving your team on the lines and not when hitched. Once hitched, the traces take over to prevent the hind ends of the horses from flaring apart. This breeching tie is about 2 feet long. A shorter breeching tie might adversely affect lateral alignment while hitched.

42 inches apart, which is what I consider about the middle of a broad range.

Often when I lecture about stub lines and the problems that occur when they are too short, I'll ask a group of teamsters if they have ever had a line hang up in a hame ring. Judging by the number of pained grins and nodding heads I've seen, this is a problem that many teamsters have experienced firsthand. Getting a line hung up usually leads to a very bad wreck. With longer stub lines, when a troubled pair of horses is recklessly seesawing back and forth on the doubletrees, the old fear of getting a stub line hung up is completely gone. Nothing they can do by jumping around can exceed what the lines were made to accommodate. Nothing they can do will cause you to completely lose control of them by getting a line hung up. It's a great feeling to solve the case of the disappearing stub line.

THE BREECHING TIE

There are few pieces of equipment more humble than the breeching tie. Usually made of cast-off material (an old lead rope, worn-out quarter strap, or piece of baler twine), it is regarded as unnecessary by many teamsters. I've had highly experienced teamsters question why I would use a breeching tie. My reply is that there is nothing about an unhitched pair of horses that keeps their hind ends together. When they are hitched, the traces attached to the doubletree keep them together at the rear, but while you are unhitched and driving out to whatever you are going to hitch to, it's a great way to prevent a problem.

Once, as a friend was about to drive out of the barn with his new team of mules, I noticed that he wasn't using a breeching tie. I spoke up right away and asked, "Don't you want some kind of a breeching tie on them?" He gave me a funny look and said that he figured they were better trained than that. I didn't say anything more as I followed him out to his wagon to help him hitch. As he approached the wagon tongue, you could see the mules were getting a little nervous about the new vehicle. They stepped over the tongue, but not without a lot of body twisting as they picked their heads up and tried to find an angle where they could keep an eye on their suspicious new surroundings.

Sure enough, he got them positioned about where he wanted them at the tongue and one of them looked over her shoulder, right at him. Before either of us could do anything, she walked straight toward him and threw the whole scene into chaos. The other mule wasn't sure what was happening, but she knew it couldn't be good, so she took off back toward the barn. Meanwhile, the first mule was dragged along by the lines. They ended up circling each other at a pretty good clip before they finally broke a line and then they both took off for the barn. As they ran into the barn he said, "I see what you mean about that breeching tie!"

SNAP POSITIONING

Here's a little-known safety subject that needs to get a little more attention. All snaps used on harness have a solid side and a spring-opening side. The solid side of the snap (some call it the blind side) is the one doing the work of linking two pieces of equipment together. The spring side of the snap is there to make it quick and easy to link and de-link parts. I often see teamsters driving around without realizing that those two sides of a snap also have the qualities of being safe and unsafe.

The safe side of a snap is the side that doesn't open. The unsafe side of the snap is the side that springs open. When given an opportunity, it is always better to use snaps so the safe side is presented toward the action. Horses, even well-trained ones in harness, move their heads and necks a lot, especially after they've been working and are then brought to a stop. Horses do a lot of stretching and rubbing to make the harness settle into a more comfortable position. Leaving anything around the working space of a horse, where he might get hooked up or caught, doesn't make sense. A simple thing like snap positioning makes it so easy to prevent most problems.

▲ **Coupler and breast strap snap pointing.** The solid (blind) side of the coupler snap should be toward the outside (lateral) side of the pair of horses. Only one breast strap snap is needed on the right side of a breast strap. Turning the snap so the blind side is facing outward prevents problems because it makes it nearly impossible to get a bit ring hung up.

When a horse is standing next to his teammate and he gets his bit fastened into the coupler snap along with the neck yoke, you've got a problem. Most horses are not so well trained regarding response to pressure that they'll just relax and stay there until you turn them loose. Usually, the panicked horse begins to lurch backward in a tight curve. The trouble is that most wagons and implements don't handle rapid backing in a tight curve very well. You'll be running into things behind you, breaking bridles, and you might even flip over because the tongue is jammed beyond its turning ability. When you see a picture of a team hitched to a vehicle, you can tell at a glance how safe the teamster is and whether he or she safely positioned the snaps.

SEEING THE WHOLE PICTURE: AN ARGUMENT FOR OPEN BRIDLES

I was talking on the phone with my nephew and mentioned that I was now driving most of our horses in open bridles and teaching people the benefits of driving that way. Immediately, he fired back at me, "Are you crazy? That's a good way to get people hurt! Aren't you the one who spent so much time teaching the benefits of blinders and how to position them?" I could see he wasn't going to accept my new way of doing things without a fight. His attitude is shared by many who have never considered using open bridles. If you want to be looked at with complete distrust and suspicion, just approach virtually any show grounds in the United States driving a horse without blinds. Most of the people watching would want to usher you out of there in a big hurry before your horse ran off and wrecked something.

Teamsters Do Use Open Bridles

After giving my nephew a moment to calm down, I started explaining my reasons for the big change from blinders to openness. It seemed to comfort him a little when I told him that I am not the only one who uses open bridles. In many foreign countries, horses are more often driven without blinds than with them. Especially in highly complex, crowded traffic situations, horses will regularly be seen working without the use of blinds.

High-Performance Open-Bridle Horses

Then I gave my nephew the historical perspective. Around the early part of the twentieth century, before the introduction of motorized fire equipment, horses were used to pull the vehicles needed to fight a fire. Virtually all of the horses used to pull fire equipment were driven without blind bridles and were under control. Imagine the embarrassment of the fire department if their charging steeds went tearing right past the building that had the smoke pouring out the windows! I've never seen a picture of a fire horse that was wearing blinds unless it is a picture of a team pulling an old pumper in a modern-day parade.

Artillery horses that pulled cannons in battle is another example of a high-performance use for open bridle horses that far exceeds what most of us are doing with our modern animals driven in blinds. These equines were trained to be ridden while driven postilion as four-up and trained so that they could be used anywhere in the hitch. Battlefield work often had to be done at a dead run and over rough ground to a precise location where the cannon was deployed. If you saw the horse artillery part of Ronald Reagan's televised funeral procession, you might have noticed that those extremely well-behaved artillery horses were being driven in open bridles.

What is the Benefit of Seeing?

I also had a philosophical argument for open bridles. No matter how you look at it, removing some to nearly all of a horse's ability to see is a restraint technique. It is not a relational restraint technique like the use of the bit in a horse's mouth because blinders are either there or they are not. I've never seen covering and uncovering of a horse's eyes being used to steer and stop horses when driving, such as what happens when a human rides an ostrich. When blinders go on prior to hitching, they are positioned carefully to stay there throughout the drive, no matter how the horse is acting. The use of blinds is not behavior dependent, which makes it a restraint device that is not relational. Horses are highly relational animals, so I think it is better to use training techniques that show these animals you are every bit as relational as they are.

▲ **No blinds.** There are good reasons to use open bridles—including showing your horses you trust them.

Going a little deeper, because blinders are restraint devices that aren't used relationally, the blinders are used as communication devices. They communicate to the horse that he is not being trusted with the full use of his faculties. This is a powerful yet subtle way of telling the horse that you want to use his body without using his mind. In other words, blinder use tells the horse that you don't trust how he would use his brain if he could fully see what you are doing to him. One of my favorite quotes on the subject of trust is, "If you introduce an element of mistrust into a relationship, it ends the communication." A big part of the language of trust is openness—not covering up things.

For many, having their horses do their jobs without mutual trust or respect for the driver is apparently desirable. One of the most basic reasons for a horse's feet to move is fear. Many teamsters place a high value on that fear factor that accompanies blinder use because it is the only way they know to get their work done. A horse will pull their load and pull hard on the lines hour after hour and day after day because of the fear of the vehicle he is pulling. Yet another way to get those feet moving is with respect. Training your horse to move forward with energy into the load for extended periods of time by using respect instead of fear tactics requires a completely different way of training. Often it seems that breaking uses fear as a motivating factor. By first removing the fear, training relies on a buildup of trust and respect to move the feet. Open bridles are more associated with the way of handling horses that starts by training instead of breaking.

Blinders and Imagination

My nephew seemed to come around to seeing things my way, so I added a new twist to the argument and told him my all-time favorite reason for going without blinds—to increase calmness. If you cover up a horse's ability to see what is happening around him, you are expanding the possibilities for him to imagine things that aren't really there. I often hear stories from teamsters whose horses do okay as they approach a yappy little dog along the side of the road, and then have a runaway as soon as the blinders hide the dog. Imagination, mistrust, and the inability to see what is actually happening can cause sudden panic in horses.

By careful relationship building during training, and desensitizing any concern about the load coming along behind, horses learn that there is nothing to fear. That is why the main characteristic of horses trained to be driven in open bridles is their completely calm and sensible way of acting. If the load coming along behind the horse isn't scaring him, he has a greatly improved mental capacity—a brain uncluttered by fear. Because he can see everything, the horse is more likely to be okay about other things, like yappy dogs, that might intrude on the working environment.

Horses that are trained to be driven in blinds will usually run off if they are suddenly allowed to see the load they are pulling. This alone should be reason enough for teamsters to get their horses out of blinds after careful training. Relying on a little piece of equipment's positioning to prevent certain catastrophe is not a great feeling while you're driving! You don't need to talk runaway stories with experienced teamsters for very long before you hear again the one about the bad runaway that started when one of the horses rubbed his blinder off. I've gotten the blinds off of lots of horses in the past few years, but every one of those horses had to be carefully trained for the new, open way of seeing things.

It's important to know that some horses seem to be happier working in blinds. When training young horses we take them in open bridles for as long as possible. Any horses that seem happier in blinds are worked that way. By doing this the horse helps decide what is best for him.

Even if you have no interest in going without blinders, you might someday encounter a teamster who is driving with open bridles. Contrary to my nephew's initial reaction, going with open bridles isn't something teamsters do because of craziness. Most teamsters who are driving their horses in open bridles are doing it because they think it makes driving much safer. Like I told him, instead of having a "blind" attitude, maybe it would be better to be open to seeing things differently!

> "What custom hath endeared
>   We part with sadly,
>   Tho we prize it not."
> —Joanna Baillie

Chapter Nine

Getting and Keeping the Right Horses

Your success or failure at farming with horses is going to depend on getting and keeping the right horse(s). If you get the right horses, putting them to work on your farm will make your days pleasant and rewarding. If you end up with the wrong kind of horses, your days will be filled with all sorts of stress and anxiety right from the start.

Buying horses privately is a great way to go if you can find a team that way. You get to hear the complete history of the horses, you can see them being worked on their home turf, and you don't have to make a snap decision about whether you want them or not. It's usually okay to come back and look again while you're deciding. The biggest disadvantage to buying horses privately is that you usually have to drive a distance to look at them, and sometimes once you get there, the horses aren't quite as good as their owner made them sound over the phone.

▲ A nice first team?

At most draft horse auctions around the country, you'll see plenty of teams and singles for sale. The trick at an auction is to make sure you're getting good, honest horses and not some dishonest horse trader's pack of lies. I usually study the person representing the horses just as much as the horses themselves. I'm not great at telling honest people from dishonest, so I usually consult with local horse traders about the seller's reputation before I believe much of the sales talk.

People buying their first team at a major sale sometimes get the wrong horses. It's very disappointing to see honest, well-meaning folks who want to use a team of good horses on their farm get roped into buying the wrong ones for the job. Those flashy-looking young horses with their heads in the air and their mouths open from pulling 60 pounds of pressure on the lines aren't going to make it as a working team. To get the right first team, you'll need to look at horses that look like they'll work with, not against, the new teamster. Horses that do their work on slack reins, with their heads down most of the time, are far better than having a team that could win the hitch class at the state fair. Usually, but not always, the best horses for your first team will be older and well used with a great reputation. Ideal beginner teams are about 12 to 15 years old and have been used for honest work all of their lives. After you've used this team on your place for a few years, you'll start to acquire the skills that will make it safer for you to buy a younger team next time.

BEING PERSISTENT

If you happen to buy the wrong horses for the work you have, that doesn't have to be the end of the story. Like the bumper sticker on the teacher's car says, "If you think education is expensive, try ignorance!" If you get the wrong horse or team the first time, think of it as a tremendous learning experience. Horses that are too much trouble should be sold to a more suitable home or traded in on what you now know you really need. Use that educational experience to start over right the next time.

When you see a great-looking hitch of horses that are being exhibited all over the country, do you assume that those horses have been together since they were colts? That almost never happens. In high-performance jobs where the behavior has to be nearly perfect, new horses are bought and difficult horses are sold all the time. It's part of the normal cost of having a successful horse business.

If you get the feeling that your problem is extreme lack of experience on your own part, then you should be prepared to educate yourself and learn. There are a growing number of workshops offered all over the country that teach people how to drive horses for farming, ranching, logging, and street use. If you know of anyone working horses or mules in your area, you might be able to get a job or at least be allowed to give free help in exchange for soaking up some culture. Even if you don't totally agree with everything they

◀ **An ideal pen.** A log-sided 60-by-60-foot pen with shelter is an ideal place to keep one to six head of horses when they can't be out on pasture. It is big enough to allow full-bore playing and the social interaction that horses love.

▶ **Horses on pasture.**
It doesn't get much better than this for horses that have time off from work. They get to do one of their favorite activities: eat grass!

do, you can still get an education about "how not to do things," which is a whole lot more than knowing nothing.

There are those who say that horse farming will never make a comeback because horse farmers are raised, not trained. I might agree with that depressing thought if I hadn't so often been impressed with the tremendous effect a person with persistence can have on a bad situation. I've seen a guy with only one arm make a living shoeing horses. I've seen a guy with only one hand drive a six-up of horses. It doesn't strike me as impossible for a person with a desire to use horses to be wonderfully successful, even with all of the difficulties caused by lack of experience. Success or failure depends mostly on the willpower of the person.

KEEPING YOUR HORSES

The all-time most acceptable way to house horses seems to be in stalls. Keeping a big horse in a stall has a lot of advantages for the owner, but none for the horse. These big animals were made to have room to move. Being stuck in a stall 24 hours a day is about the same as putting a human in a prison cell the size of the average hall closet. The wall or door is in their face all the time. Horses can't run and play in a stall, so they can't get exercise as loose horses do. Lack of movement and standing in excrement causes disease in the horses' feet.

Along with being confined, another thing that is hard on horses is the common practice of grain diets. It's usually cheaper and easier for the horse owner to feed grain and a little alfalfa hay instead of giving the horse free access to grass hay. The horse will wolf down his high energy morning ration of alfalfa and grain in about 15 minutes and then spends the next eight hours waiting for supper. Breakfast is a 16-hour wait!

Horses were made to have free access to grass in a natural environment. In my opinion, the nicest way to keep horses is to at least have them in pens that are big enough for them to run and play if they feel like it. It's also nice to give your horses grass hay in abundant quantities whenever grazing grass is unavailable because eating all day is an activity that keeps horses nicely occupied. If horses are overfed a concentrated ration and they can't move around, when you do have some work for them to do, it may not go so well. Horses are like humans. Too much confinement in an unnatural environment followed by freedom makes them act bonkers. However, if your horses have a real job to do for more than a couple of hours each day, then confinement and feeding grain during the off hours is not detrimental to their mental health.

GETTING THE RIGHT LOAD

Hitching to a load that is more than your horse was ready to pull can make your horse balky. Balky horses have a phobia about pulling that wrecks their confidence to pull loads. When hitched, balky horses refuse to go forward. Even heavy whip use is completely ignored by a balky horse. If pressured severely, a balky horse will suddenly plunge forward, but return to the "stuck" condition the next time he is stopped. After the horse is hooked to a load that is too heavy, balkiness will begin to rear its ugly head. Balkiness is a serious mental illness that is usually not easily fixed.

The best remedy for the balkiness problem is to never let it develop in the first place. Experienced teamsters know if a load is too much weight for a horse to pull, and they stay on the prudent side when selecting loads. At minimum, a well-harnessed, physically comfortable horse should be able to draw horizontally 10 percent of his body weight for extended periods of time. For short bursts of extreme effort, you can expect a horse to pull up to

100 percent of what he weighs and more. Getting such high levels of pulling energy requires in-depth knowledge of collar fit considerations and careful physical conditioning over time, coupled with careful mental training for confidence with a load. Horses can pull incredible loads when the teamster knows how to inspire unyielding confidence in his horses.

Another form of overloading that can happen, even when the load is at 10 percent or less of body weight, is loading caused by duration of pull. Especially on a hot, humid day, working horses need to be checked often to be certain they are not being overworked. It's a good idea to stop every 10 minutes as the work is started in the morning to see what effect your load is having on respiration. If you work horses for a long duration where their respiration rate is higher than their heart rate, they can become overstressed and require quite a while to recover from such treatment. I like to check often to be sure that the respiration rate stays below 60 breaths per minute, which is one breath per second. You can easily see the respiration of working horses from the implement seat. The remedy for overwork is rest, and it is best to take a lot of little rest stops. That resting practice will keep you in the field for longer periods of time and your conditioned horses will be more ready to go to the job site each day.

Just as it is possible to overload horses, it is also possible to underload them. Horses can get too revved up by not ever having to really do work. If you usually only get your team out to hook them onto a forecart or an empty wagon, you will soon begin to see problems. On such light loads, often the teamster is the one doing the actual pulling of the load through the lines. As he or she leans back on the lines to restrain the horses' energy, there is enough tension in the lines to shove the load forward with the driver's feet, which are pushing against the floorboards of the vehicle. If you look at the traces, you'll see that they are slack, even going up a hill. Another problem with light loads is the way the slack traces expose the horses to being bumped by the pole. Traces can only protect horses from slamming into the pole while turning if they have a certain amount of tension in them. Those sorts of loads, if used exclusively, will quickly make a nice pair of horses start acting like idiots.

KEEP THE NOISE

If you don't have work that is real and necessary with your new team, it is best to hitch onto something like a sled. Training sleds on dirt and gravel with a person or two on the sled usually pull about right. This sort of load pulls hard enough to keep a team moderately worked, but is not so light that the horses feel like they are playing every time they are out of the pasture/corral. Sleds make a lot of noise when driven over gravel. Noise is good. If you only ever hook onto silent, rubber-tired vehicles, your horses might forget they can tolerate noise in the working environment. This conditioning helps your horses handle the noise of heavy traffic, unexpected bumping and thumping of normal load hauling, and the machinery noise of some horse-drawn implements.

THINK ABOUT MENTAL PREPARATION

Probably the biggest thing new teamsters need to think about is how well their horses are mentally prepared for doing the job at hand. If you have full-time, all-day projects for your new team of horses, you won't need to think about preparing your horses mentally for a task. If you bought a nice team of horses, but you've been tied up at your other job for the past two weeks while the horses have been kept in stalls and fed grain to fatten them up, you might be in for a surprise at the difference in attitude your horses are now showing. Underworked, overfed horses should be reintroduced to being cooperative before you hitch them onto a manure spreader and turn the thing on. It would be better to hitch onto a sled for a half-hour or so at a walk to burn off excess energy before hitching them onto the manure spreader.

Many old-time teamsters made it a practice to rarely trot their draft horses. Many modern-day teamsters seem to spend more time trotting than walking. In most cases, your horses will stay much calmer if you stick to a walk. If you absolutely must trot, only do so when you intuitively know that trotting is not likely to get the horses fired up. Going up a hill is one place where people like to let their horses trot, but a little bit of that can make horses want to charge.

You can use that to advantage on horses that need to perk up a little, but if you want to calm a team down, keep them walking when they want to trot. I've seen new teamsters change calm teams into idiots when they misuse trotting.

HORSES LIKE HIGH-QUALITY INTERACTION

You know what happens if you leave your dog at home when he's used to going along: you come home to a mess, usually on the carpet. Horses get that way when you don't have a real job for them to do. They were made to have a great relationship with you and help you do a job with a purpose, and they don't take too kindly to being ignored. Instead of getting your attention with the carpet, horses that are being left out of a relationship get highly energized, act spooky, and are belligerent about doing anything useful. Horses were made to be treated better than that, and often wild behavior is simply their way of saying, "I need to be treated better." The best horse people realize that owning a horse is making a commitment and they arrange their schedule so that their horses don't feel left out.

If your "best-case scenario" before you buy a horse or team is that you might be able to use the animals about every third weekend for one day during the spring and summer months, then please don't buy horses. It would be better if you let someone else who has a real job for them buy the horses. Most horse owners have time to make sure their horses are fed, but if you propose to actually get meaningful work out of your horses, it's going to take more than that. Horses that are great to work with won't stay that way if you don't need them.

Horses are amazing animals because of how nice they act when they have a real job to do and their owner knows how to make that job pleasant. There is an attractive look to horses that are good at their job that you will never see in horses that are petted and played with. I hope that this book helps you to become the kind of horse person who knows what's involved in using your horses for real work, and that it inspires confidence to get your horses put to work on a job.

▲ **Team on a sled.**

Index

About the Author

Steve Bowers
August 24, 1954 – June 1, 2007

Steve and his twin brother, Mike, were born in Thurmont, MD. They were raised on a horse and cattle farm and grew up working with animals. The "boys," as people fondly referred to them even when they were men, started driving ponies and riding horses at a very early age. Steve and Mike moved to Colorado in the 1970s to attend Colorado State University to become veterinarians. In no time at all he and Mike were too busy to attend school because they had their own horse training business. The "boys" galloped Thoroughbreds, trained for Arabian breeders, and were well known for training tough horses that no one else wanted to train. While they worked for Corky Keen galloping Thoroughbreds, they also began driving his draft horses, and Steve's love of driving bloomed. The brothers expanded their business and began giving hayrides and sleigh rides, and in 1979 gave their first driving clinic. The brothers also logged and farmed with horses.

Steve married Peggy in 1982 and they had two children: Katie and Nate. In the early 1990s Mike moved to Virginia. Steve continued training horses and he and Peggy also farmed with horses—their kids grew up sitting behind a team going around a hay field.

Steve had a passion for horses and wanted to help people understand them and get along with them. He decided to refocus his business on training both the horses and the owners, and he started an intensified stay at the farm where clients could bring their horse and spend the week taking lessons.

Steve loved to write and could be found many evenings working on articles. He wrote articles for the *The Draft Horse Journal*, *Driving Digest*, and other horse magazines. He published two books and produced three videos.

Steve died suddenly in 2007 from a brain aneurism. The following was a quote from a card sent after his death:

"When Steve died the world lost a wonderful man and a truly great horseman."

Also available from Voyageur Press

How To Raise Chickens
(ISBN 978-0-7603-4377-7)

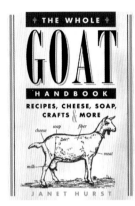

The Whole Goat Handbook
(ISBN 978-0-7603-4236-7)

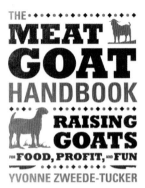

The Meat Goat Handbook
(ISBN 978-0-7603-4042-4)

The Beginner's Guide
to Beekeeping
(ISBN 978-0-7603-4447-7)

Homemade Cheese
(ISBN 978-0-7603-3848-3)

The Complete Book of Butchering,
Smoking, Curing,
and Sausage Making
(ISBN 978-0-7603-3782-0)